Little Acorn Books

Readiness Games
35 Folder Games For Numbers

by Marilynn G. Barr

LAB20145P
Readiness Games
35 FOLDERGAMES FOR NUMBERS
Preschool — Grade 1
(*Skills Focus: readiness skills, recognizing numerals and number words, early math concepts, counting, identifying number sets, telling time, sequencing, hands-on fine motor skills responses, following directions, fair play*)

by Marilynn G. Barr

Published by: Little Acorn Books™
Originally published by: Monday Morning Books, Inc.

Entire contents copyright © 2014 Little Acorn Books™

Little Acorn Books
PO Box 8787
Greensboro, NC 27419-0787

Promoting Early Skills for a Lifetime™

Little Acorn Books™
is an imprint of Little Acorn Associates, Inc.

http://www.littleacornbooks.com

Permission is hereby granted to reproduce student materials in this book for non-commercial individual or classroom use. *School-wide or system-wide use is expressly prohibited.

ISBN 978-1-937257-53-8

Printed in the United States of America

35 FolderGames for Numbers
Contents

Introduction 4
  General Directions 4
  Activity Use 4
Playing Directions
  Trail Game Playing Directions 5
  Match Board Playing Directions 6
  Dominoes Playing Directions 7
  Pocket Game Playing Directions 8
  Bingo Playing Directions 9
  Lotto Playing Directions 10
  Clothespin Match-ups Playing Directions 11
  Clothespin Wheel Playing Directions 12
  Fill-ins Playing Directions 13
  Dot-to-Dot Playing Directions 14
  Concentration Playing Directions 15
Counting Number Sets 1 to 10
  It's Tea Time 16
  Candy Heart Dominoes 21
  Turtle Shell Lotto 25
  Peek-a-boo Pandas 30
Numeral Recognition 1 to 10
  Busy Bees 35
  Firefly Fiesta 40
  Butterfly Wings 45
  A Spider's Adventure 50
Numerical Order 1 to 10
  Peek-a-boo Monkeys 55
  Polka-Dot Turtles 60
  Poppin' Popcorn 65
  Marvelous Mouse 69
  Fabulous Fish 74
Number Word Recognition 1 to 10
  Cheese and Crackers 79
  Panda Lotto 83
  Tumbling Turtles 88
  Koala King Bingo 93
  Rubber Duck Dominoes 99
  Cupcake Lotto 104
  Ladybugs and Leaves 109

Counting Number Sets 1 to 12
  Rainbow Eggs Baskets 114
Numeral Recognition 1 and 12
  Firefly Lights 119
  Polka-Dot Ponies Lotto 124
  Carnival Clown Bingo 130
  Turtle Dominoes 136
Numerical Order 1 to 12
  Zebra Zippers 140
  Great Goose 145
  Peek-a-boo Penguins 151
Number Word Recognition 1 to 12
  Peek-a-boo Roos 156
  Dancing Dragons 160
Telling Time
  Peek-a-boo Fish 165
  Sneaky Snakes 171
Early Math
  Frogs on Logs 176
  Shuffle Board Sharks 180
Counting to 100
  Whistling Whales 184
Whale Counters 190
Folder Handles 192

Introduction

FolderGames for Numbers presents activities to enrich beginning math skills for young learners in easy-to-make-and-use file folder set-ups. The folders can be used with individual children, small cooperative groups, in learning centers, or with families at home.

The activities in FolderGames for Numbers help to motivate and strengthen early math concepts and skills in an enjoyable and stimulating format. Games assist children in mastering basic math skills. The activities focus on helping children to understand the relationship between number sets and numerals, to work on time-telling skills, and to count from small quantities up to 100. Children will practice recognizing number words, counting number sets, and sequencing number sets and numerals. A variety of hands-on responses, including placing objects, clipping on clothespins, and connecting the dots, keep the children actively engaged.

Each FolderGames activity includes the file folder layout and the activity to be duplicated, easy directions for assembly, and simple directions for use.

General Directions

Use sturdy colored file folders for the FolderGames folders. Duplicate the inside file folder set-up, the illustration for the folder cover, and any game pieces. Color all parts with felt pens, colored pencils, or crayons. Then trim and cut out. Glue the file folder set-up to the inside of the folder and the illustration to the outside front. Glue any loose patterns, such as markers or cards, onto oak tag for extra-sturdiness. Color and laminate. Cut out or trim as necessary to complete construction. Buttons or other small objects may be used for markers that are not provided. Glue a manila envelope with a clasp to the back of the filer folder for games that include loose parts. Some games, such as Butterfly Wings, include non-matching cards. Non-matching cards challenge children to differentiate matches from non-matches. Patterns throughout the book can have multiple uses. For instance, make Concentration card decks using game card patterns.

Activity Use

Have the children take out any loose parts from the envelope and open the file folder on the work area. Instruct the children on how to play the game. Have the children replace any loose parts in the envelope after play. Store the folders in a file basket. Pages 5-15 provide additional information for particular activities.

Trail Game Playing Directions for
Firefly Fiesta • A Spider's Adventure • Tumbling Turtles
Firefly Lights • Sneaky Snakes • Dancing Dragons

Reproduce, cut out, and glue these directions to the back of each Trail Game folder.

Trail Game Playing Directions

Trail Games are designed for two to four players. Set up the game board on a table. Shuffle and place the game cards, face down, next to the game board. (Players may take turns shuffling game cards.) Each player, in turn, draws a card to determine where to move on the trail. Players place used cards in a discard pile. When all cards have been drawn, reshuffle the discard pile to continue playing. Play continues until each player reaches The End.

© 2014 Little Acorn Books™

Match Board Playing Directions for
It's Tea Time • Ladybugs and Leaves

Reproduce, cut out, and glue these directions to the back of each Match Board folder.

Match Board Playing Directions

Match Boards are designed for one to two players. Each player chooses one half of the board to play. Set up a match board on a table. Shuffle and place the game cards, face down, in the center of the table. (Players may take turns shuffling game cards.) Each player, in turn, draws a card. If there is a match, the player identifies the match, then places the card on his or her board. If there is no match, the player places the card, face down, in a discard pile. Play continues until each player has placed a matching card on each space on his or her match board.

© 2014 Little Acorn Books™

Dominoes Playing Directions for
Candy Heart Dominoes • Rubber Duck Dominoes • Turtle Dominoes

Reproduce, cut out, and glue these directions to the back of each Dominoes game folder. Dominoes can contain identical and mismatched number words, number sets, or numerals.

Dominoes Playing Directions

Dominoes are designed for up to four players. Shuffle and deal five dominoes to each player, then place the remaining dominoes, face down, in the center of the table. Players hold their dominoes like playing cards and do not allow the other players to see their hands. The dealer begins the game by placing one of his or her dominoes, face up, on the table. The next player looks at his or her dominoes to see if there is a match. If there is a match, the player places the matching side of his or her domino next to the domino on the table. A match can be number to number, number to number word, number to number set, and so on. If there is no match, the player draws a domino from the center of the table until he or she finds a match. Children can only play matching dominoes at the open ends of the domino structure. Play continues until no more matches can be made.

© 2014 Little Acorn Books™

Pocket Game Playing Directions for
Peek-a-boo Pandas • Polka-Dot Turtles • Peek-a-boo Monkeys
Peek-a-boo Penguins • Peek-a-boo Roos • Peek-a-boo Fish

Reproduce, cut out, and glue these directions to the back of each Pocket Game folder.

Pocket Game Playing Directions

Pocket Games are designed for one to two players. Set up a pocket game folder on a table. Take out, shuffle, and place the game cards, face down, in the center of the table. (Players may take turns shuffling game cards.) Each player, in turn, draws a card. If there is a match, the player identifies the match, then slips the card into the correct pocket. If there is no match, the player places the card, face down, in a discard pile. Play continues until all the pockets are filled.

© 2014 Little Acorn Books™

Bingo Playing Directions for
Koala King Bingo • Carnival Clown Bingo

Reproduce, cut out, and glue these directions to the back of each Bingo folder.

Bingo Playing Directions

Bingo games are designed for one to four players. Set up Bingo boards on a table. Shuffle and place the game cards, face down, in the center of the table. (Players may take turns shuffling game cards.) One player draws a card, identifies the letter, and counts the dots out loud. Each player looks on his or her Bingo board for a match. If there is a match, each player places a token on the matching space on his or her Bingo board. The first player to place five tokens (horizontally, vertically, or diagonally) on his or her board, calls out BINGO. Play continues until each player makes a BINGO. Note: Called cards are placed in a discard pile.

© 2014 Little Acorn Books™

Lotto Playing Directions for
Turtle Shell Lotto • Panda Lotto • Cupcake Lotto • Polka-Dot Ponies Lotto

Reproduce, cut out, and glue these directions to the back of each Lotto folder.

Lotto Playing Directions

Lotto games are designed for one to four players. Set up Lotto boards on a table. Shuffle and place the game cards, face down, in the center of the table. (Players may take turns shuffling game cards.) Each player, in turn, draws a card. If there is a match, the player identifies the match, then places a token on the matching space on his or her Lotto board. If there is no match, the player places the card, face down, in a discard pile. Play continues until each player has placed a token on each space on his or her Lotto board. Reshuffle deck as needed.

© 2014 Little Acorn Books™

Clothespin Match-ups Playing Directions for
Butterfly Wings • Zebra Zippers

Reproduce, cut out, and glue these directions to the back of each Clothespin Match-ups folder.

Clothespin Match-ups Playing Directions

Clothespin Match-ups are designed for one or two players. Set up a game folder on a table. Take out and place the clothespins, face down, on the table. Each player, in turn, draws a clothespin. If there is a match, the player identifies the match, then clips the clothespin to the correct space on the game folder. If there is no match, the player places the clothespin back on the table. Play continues until each space on the game folder is clipped.

© 2014 Little Acorn Books™

Clothespin Wheel Playing Directions for
Cheese and Crackers • Rainbow Eggs Basket

Reproduce, cut out, and glue these directions to the back of each Clothespin Wheel folder. Use alternate directions for Poppin' Popcorn (p. 65).

Clothespin Wheel Playing Directions

Clothespin Wheels are designed for one or two players. Set up a game wheel on a table. Take out and place the clothespins, face down, on the table. Each player, in turn, draws a clothespin. If there is a match, the player identifies the match, then clips the clothespin to the correct space on the wheel. If there is no match, the player places the clothespin back on the table. Play continues until each space on the wheel is clipped.

© 2014 Little Acorn Books™

Fill-ins Playing Directions for
Busy Bees • Frogs on Logs • Shuffle Board Sharks

Reproduce, cut out, and glue these directions to the back of each Fill-ins folder.

Fill-ins Playing Directions

Fill-ins are designed for individual play. Set up the game folder on a table. Read the directions on the game folder. Use a wipe-off crayon or marker to fill in the blanks. Have your work checked. Wipe off the folder with a soft rag or tissue when done.

Dot-to-Dot Playing Directions for
Marvelous Mouse • Fabulous Fish • Great Goose

 Reproduce, cut out, and glue these directions to the back of each Dot-to-Dot folder.

Dot-to-Dot Playing Directions

Dot-to-Dots are designed for individual play. Set up the game folder on a table. Read the directions on the game folder. Place the cards on the board one at a time. Use a wipe-off crayon or marker to connect the dots. Wipe off the folder with a soft rag or tissue when done.

© 2014 Little Acorn Books™

Concentration Playing Directions for

Reproduce, cut out, and glue these directions to the back of each Card Game folder. Create card decks with the cards found throughout this book.

Card Games Playing Directions

Two to four players can play Concentration. Shuffle and place all the cards, face down, in the center of a table. Each player, in turn, turns over two cards. If the cards match, the player takes the cards and the next player takes a turn. If the cards do not match, the player turns each card over and the next player takes a turn. Play continues until all the cards are taken.

© 2014 Little Acorn Books™

It's Tea Time
Counting Number Sets 1-10

Match Board

Assembly

Reproduce, cut out, and glue a set of oak tag handles (p. 192) to a standard-size file folder. Option: Decorate each handle with a copy of the tea cups on this page. Reproduce, color, and cut out the "It's Tea Time" patterns. Glue the match boards to the inside of the folder. Glue the cover art to the front of the folder. Decorate the border around the match boards, then laminate. Reproduce, color, laminate, then cut apart three sets of game cards. Glue an envelope to the back of the folder for game card storage. Option: Reproduce, cut out, and glue the Match Board Playing Directions to the back of the folder.

Directions
Reproduce, cut out, and glue the direction strip to the game board.

Place each tea cup on the tea pot with the matching number of sugar cubes.

16 LAB20145P • 35 FOLDERGAMES FOR NUMBERS • 978-1-937257-53-8 • © 2014 Little Acorn Books™

It's Tea Time
Counting Number Sets 1-10

Match Board

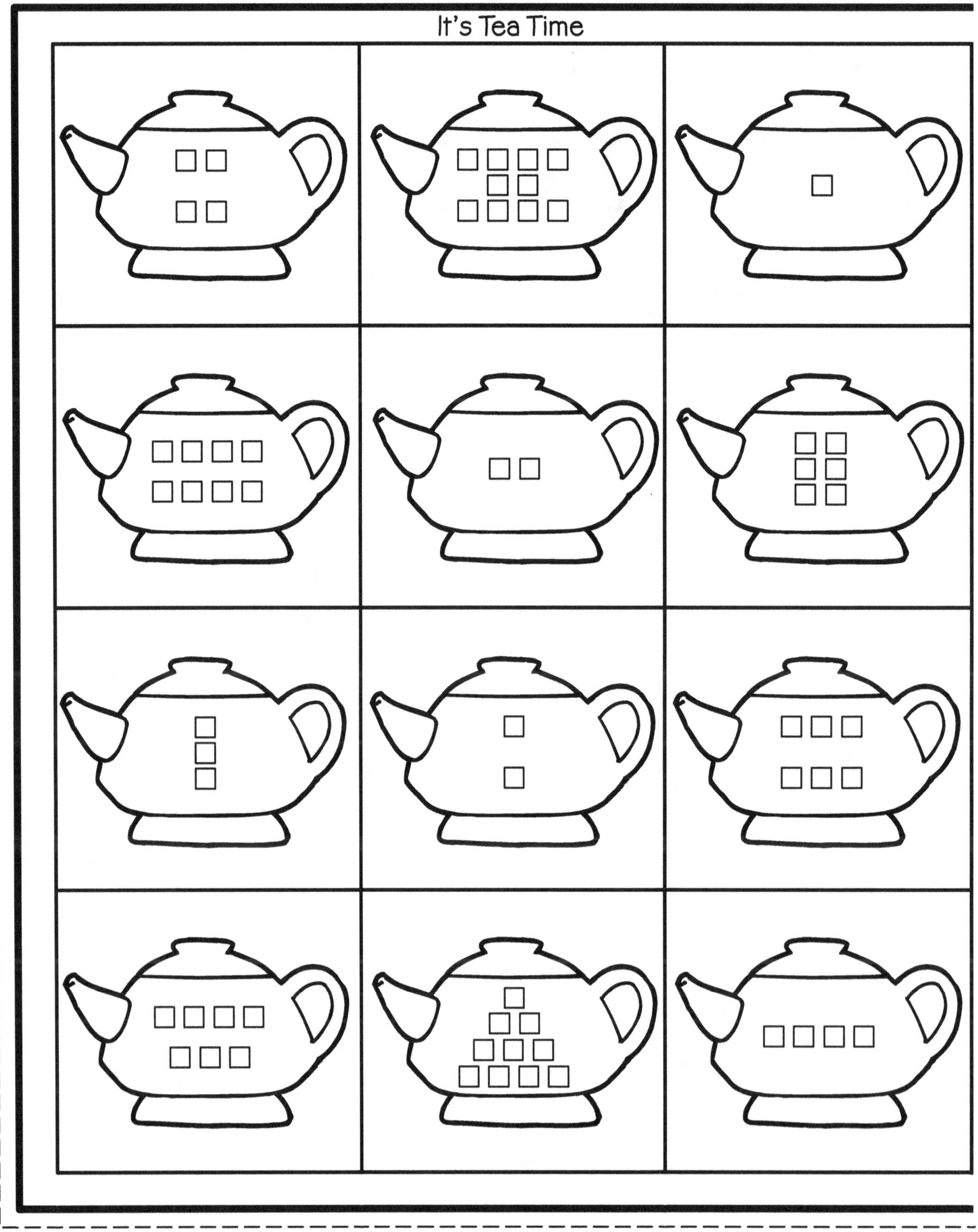

It's Tea Time

It's Tea Time
Counting Number Sets 1-10

Match Board

It's Tea Time

It's Tea Time
Counting Number Sets 1-10

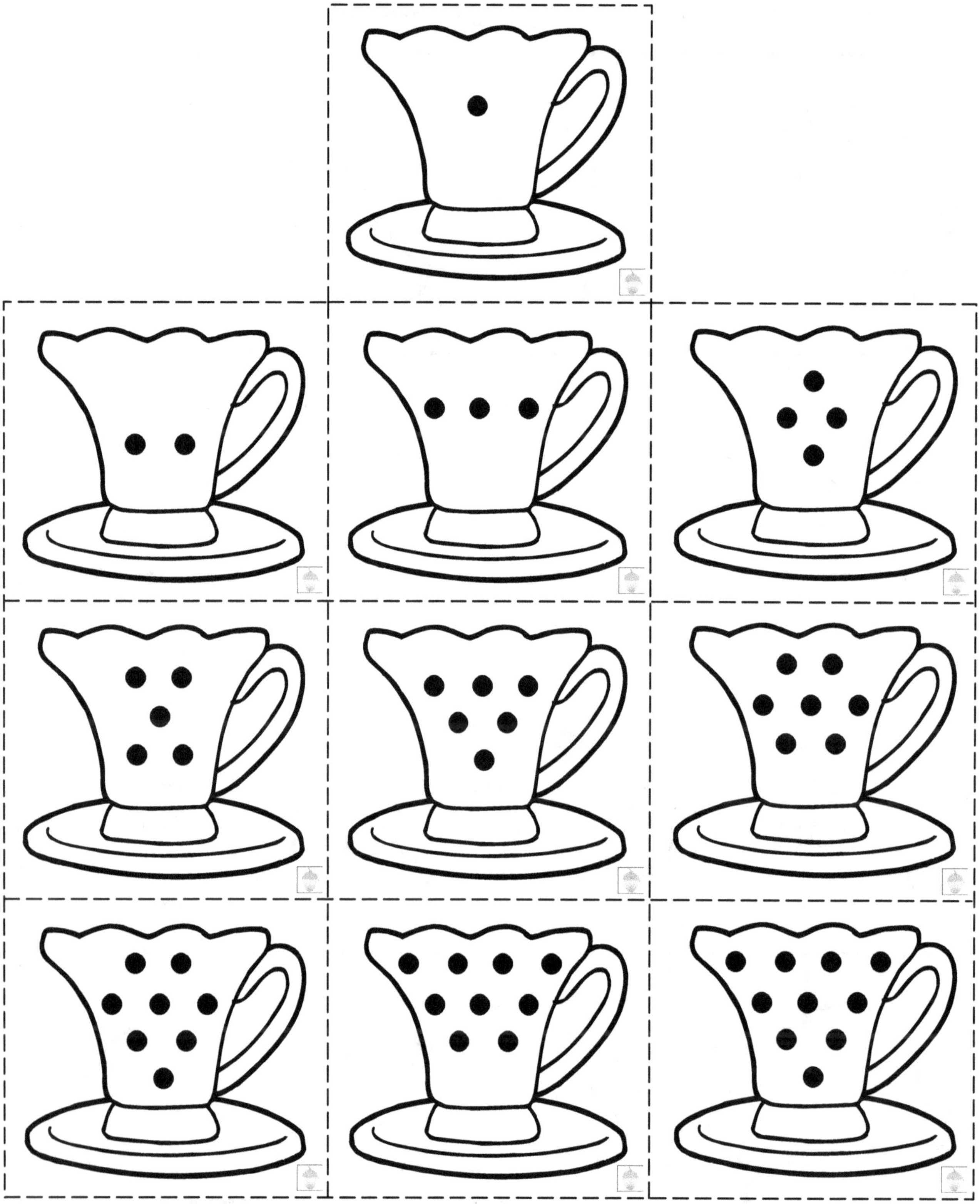

Match Board

Reproduce, color, and cut apart three sets of cards.

Candy Heart Dominoes
Counting Number Sets 1-10

Dominoes

Assembly

Reproduce, cut out, and glue a set of oak tag handles (p. 192) to a standard-size file folder. Option: Decorate the handles with copies of the candy hearts shown on this page. Reproduce, color, and cut out the "Candy Hearts" patterns. Make extra sets of dominoes to accommodate more players. Glue the cover art to the front of the folder. Decorate the inside of the folder, then laminate. Reproduce, color, laminate, then cut apart the dominoes. Glue an envelope to the back of the folder to store dominoes. Option: Reproduce, cut out, and glue the Dominoes Playing Directions to the back of the folder.

Candy Heart Dominoes Cover
Counting Number Sets 1-10

Candy Heart Dominoes
Counting Number Sets 1-10

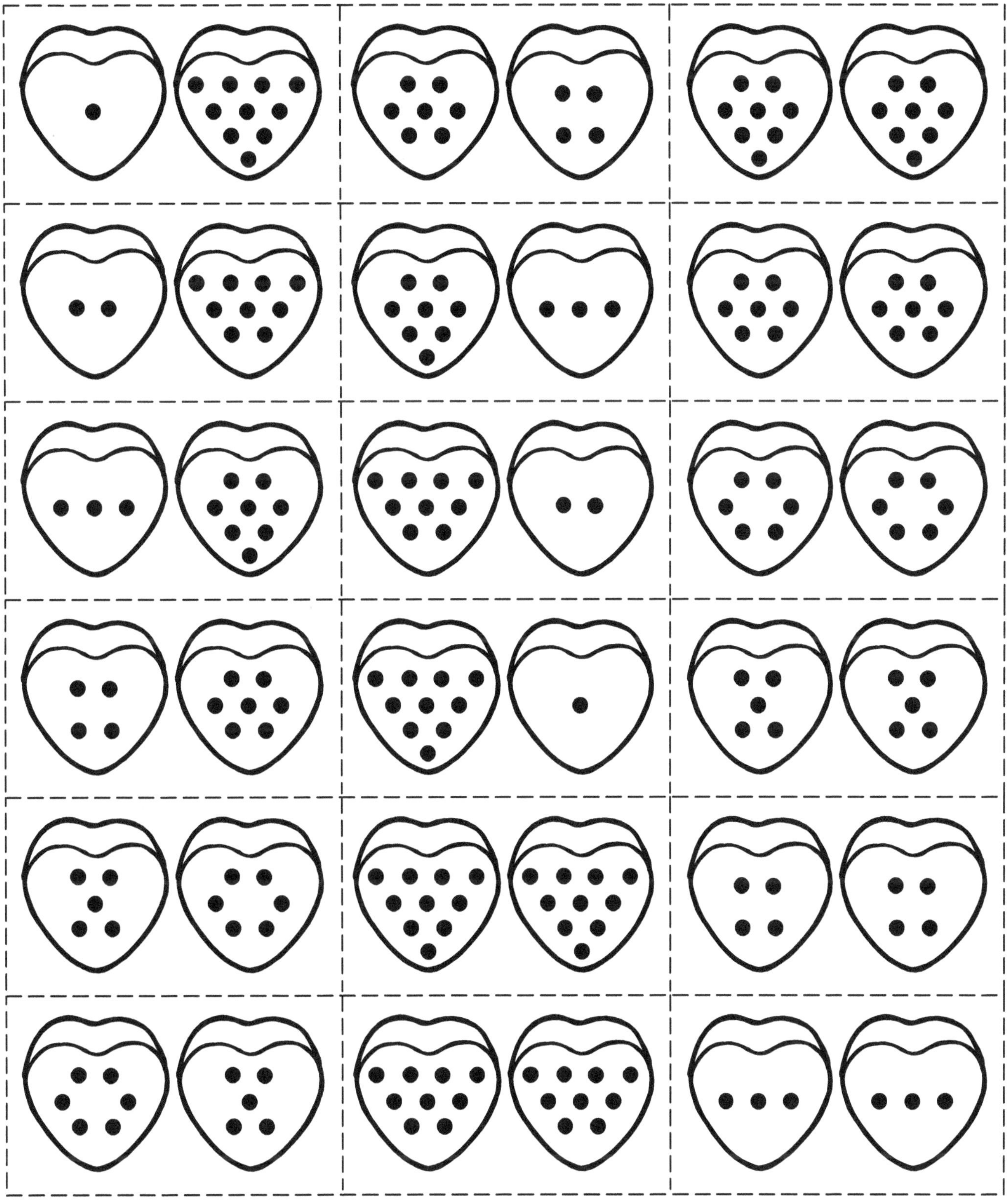

Dominoes

Reproduce, color, and cut apart.

Candy Heart Dominoes
Counting Number Sets 1-10

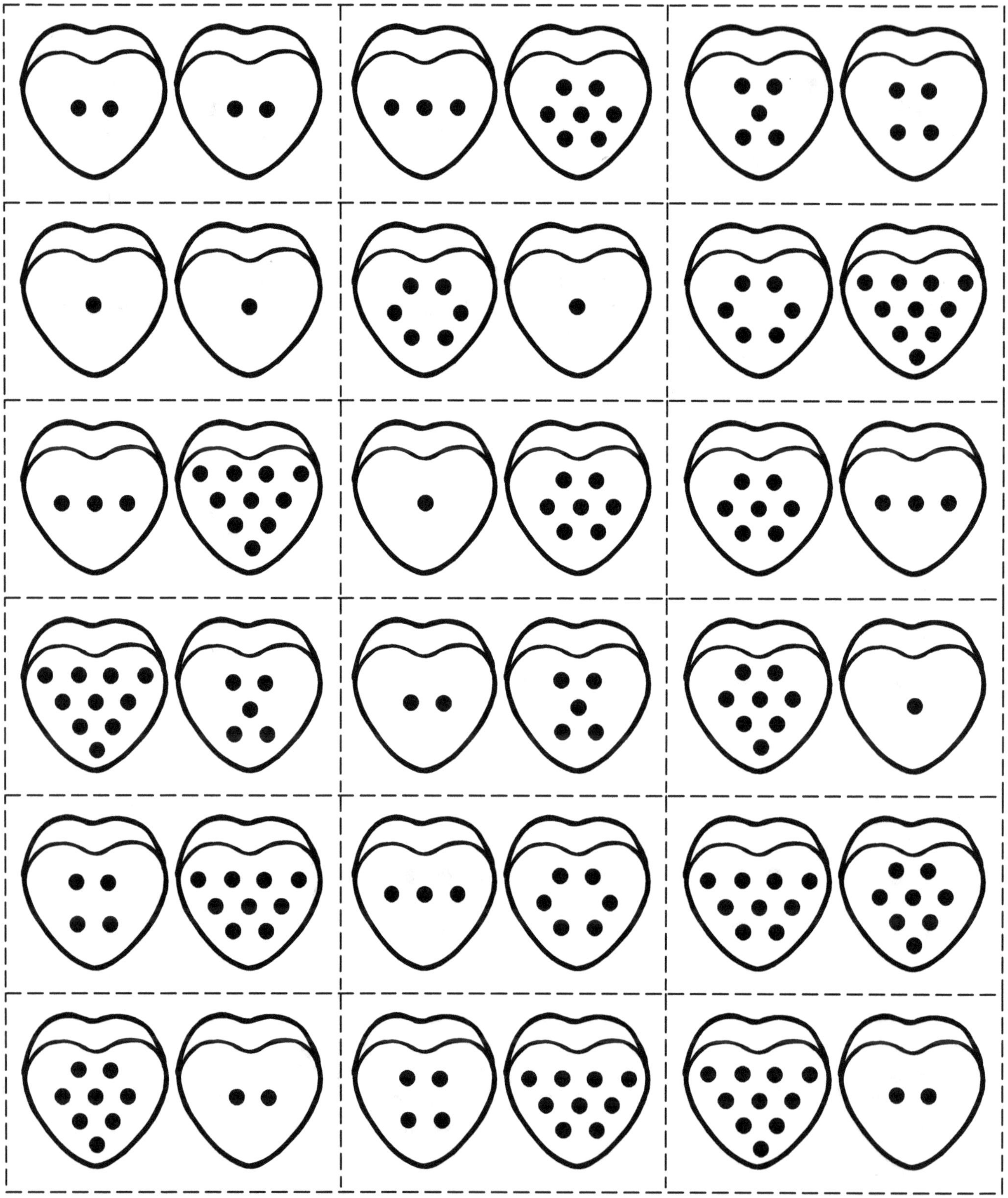

Dominoes

Reproduce, color, and cut apart.

Turtle Shell Lotto
Counting Number Sets 1-10

Lotto

Assembly

Reproduce, cut out, and glue a set of oak tag handles (p. 192) to a standard-size file folder. Option: Decorate each handle with a copy of the turtles shown above. Reproduce, color, and cut out the "Turtle Shell Lotto" patterns. Glue the cover art to the front of the folder. Mount Lotto boards on colored construction paper, then laminate. Glue two envelopes to the inside of the folder for Lotto board, token, and card storage. Option: Reproduce, cut out, and glue the Lotto Playing Directions to the back of the folder. Note: Wipe-off markers or crayons can also be used to color in matches.

Tokens

Reproduce, color, and cut out tokens for players to place on Lotto boards.

Turtle Shell Lotto Cover
Counting Number Sets 1-10

Lotto

Turtle Shell Lotto
Counting Number Sets 1-10

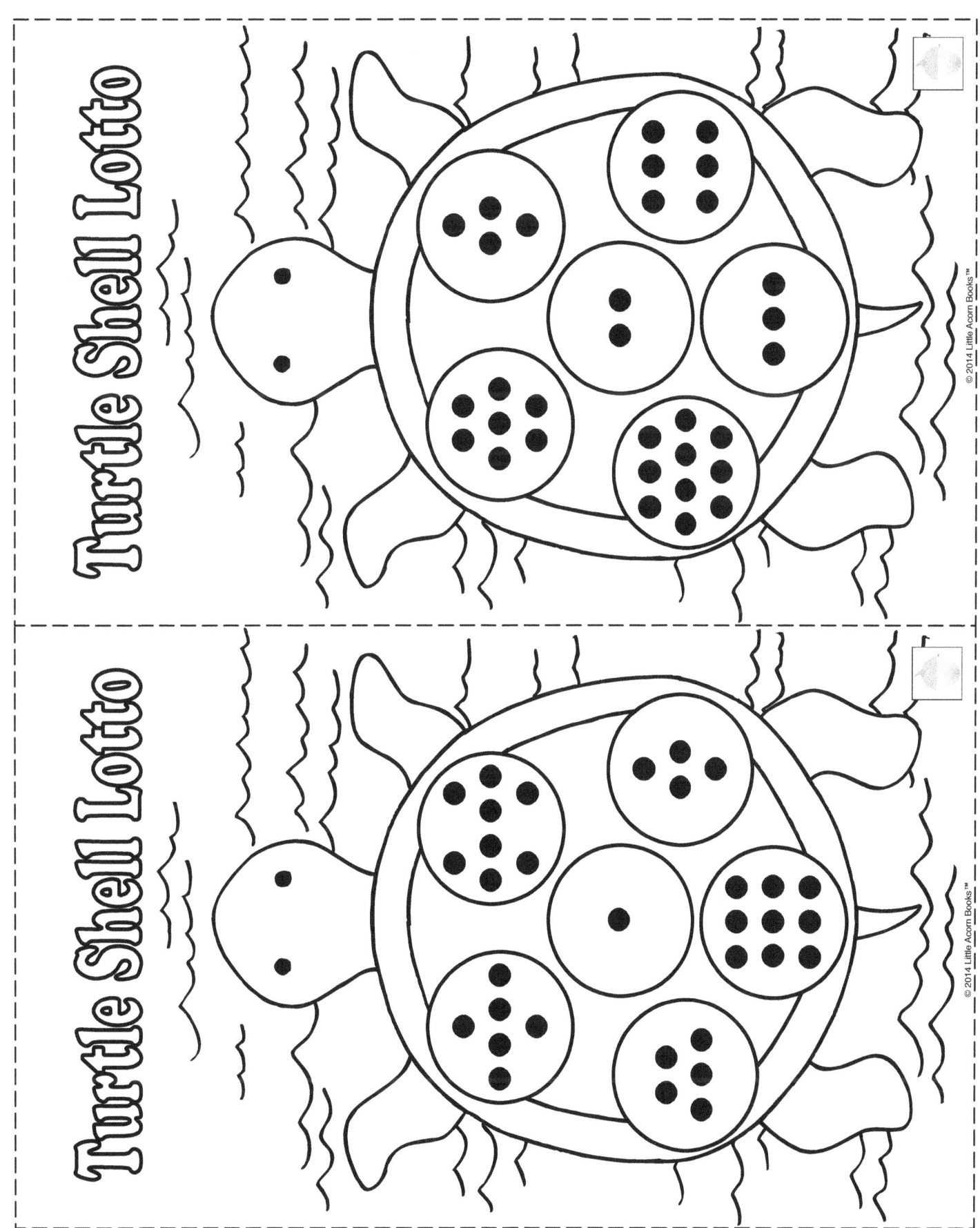

Turtle Shell Lotto
Counting Number Sets 1-10

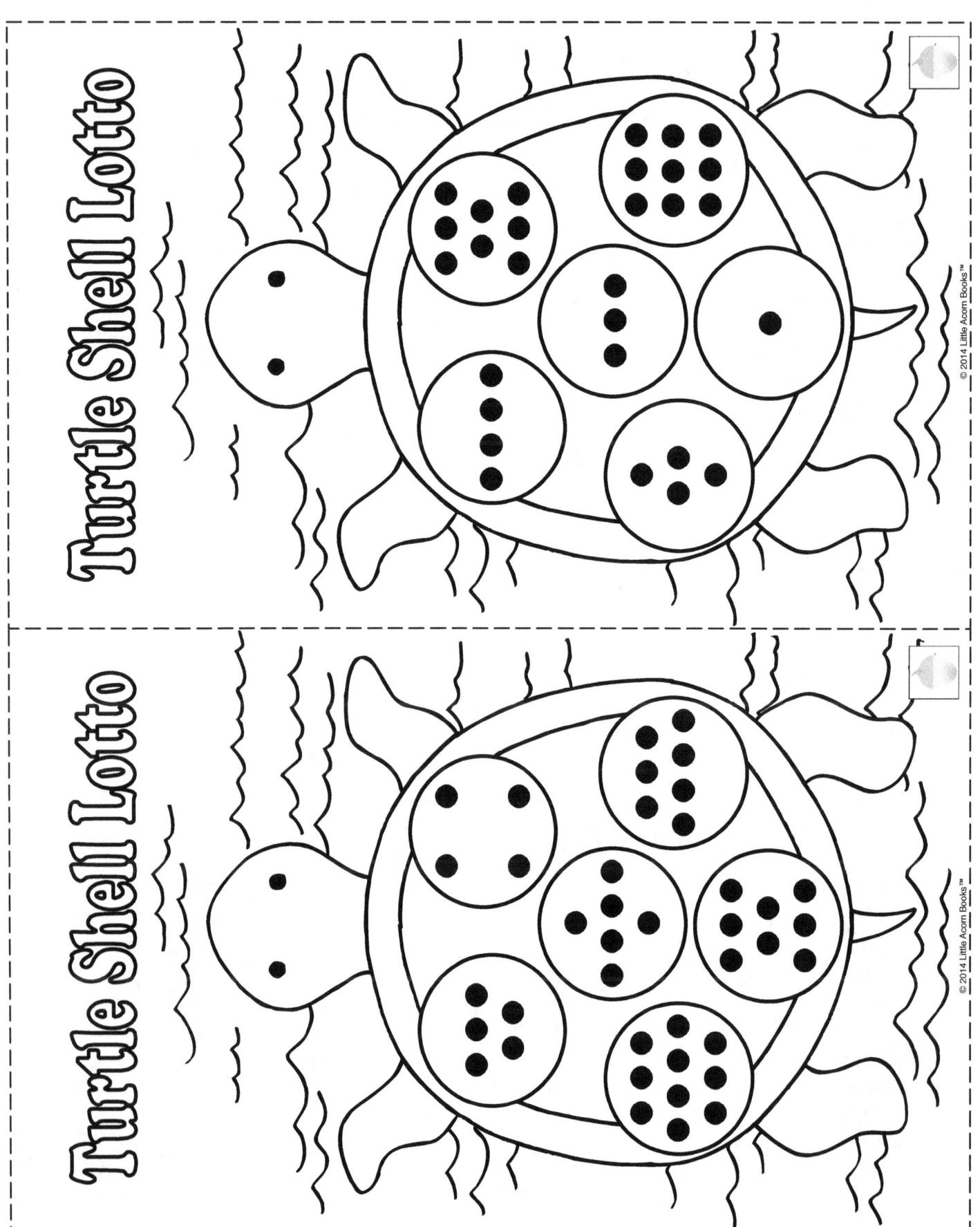

Lotto

Turtle Shell Lotto
Counting Number Sets 1-10

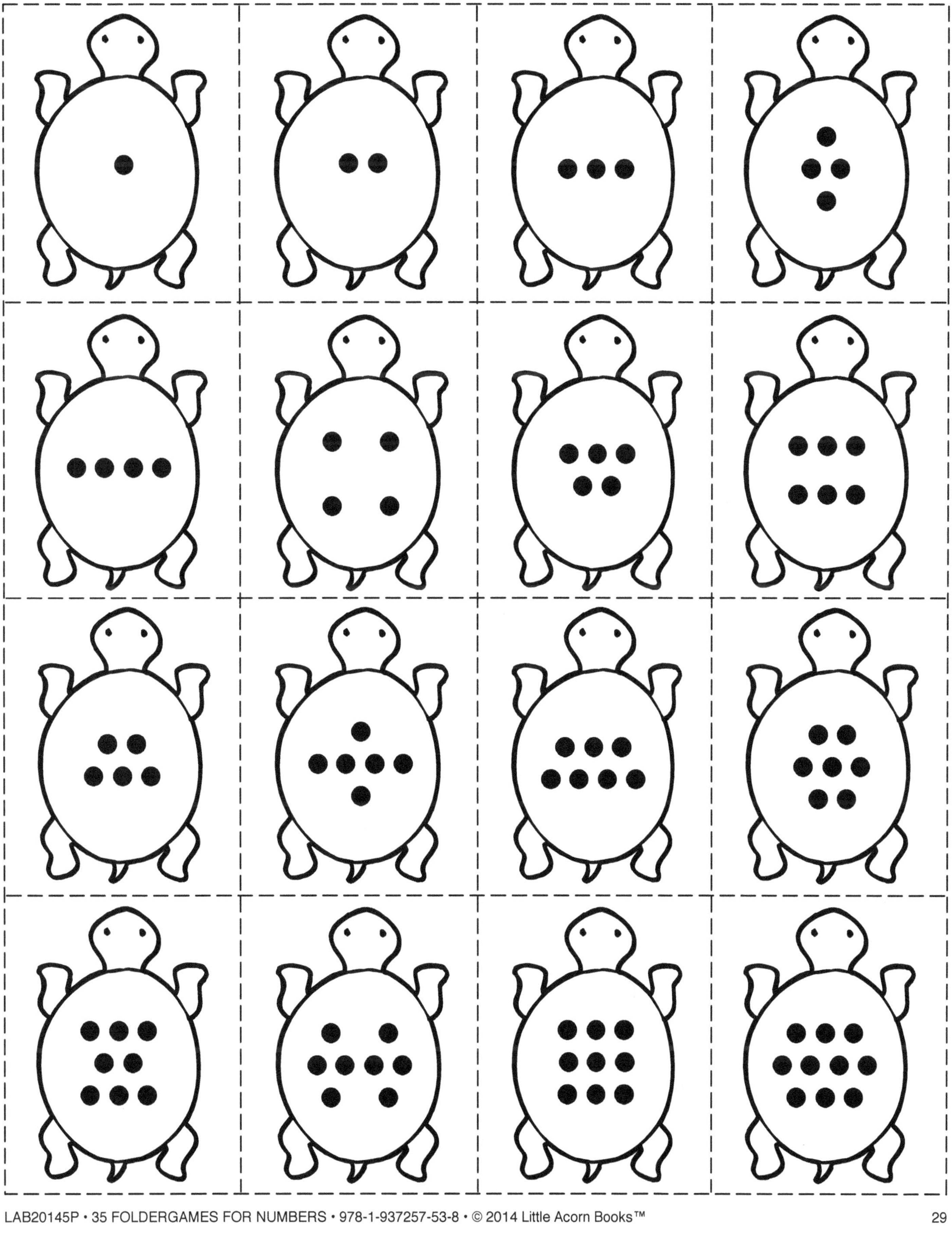

Lotto

Peek-a-boo Pandas

Pocket Game

Counting Number Sets 1-10

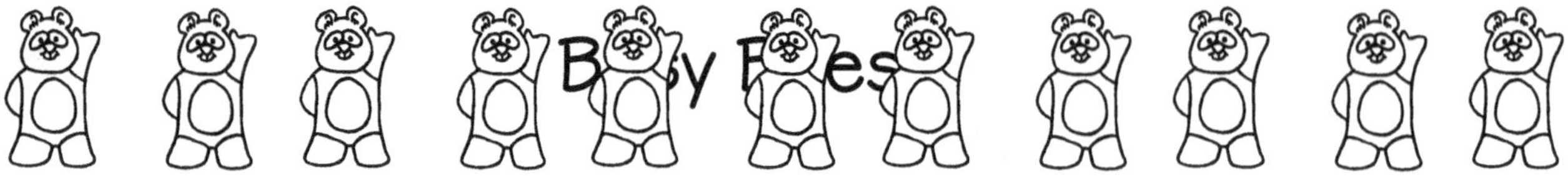

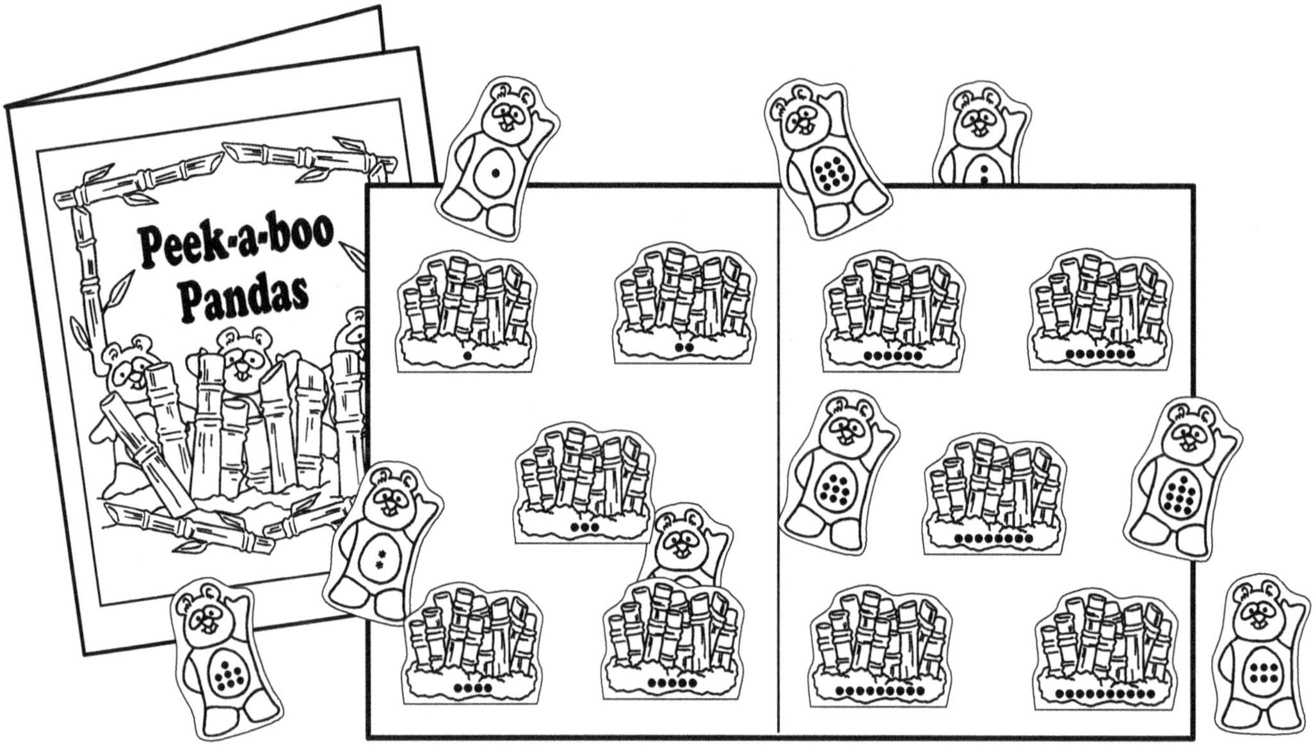

Assembly

Reproduce, cut out, and glue a set of oak tag handles (p. 192) to a standard-size file folder. Option: Decorate each handle with a copy of the pandas on this page. Reproduce, color, and cut out the "Peek-a-boo Pandas" patterns. Apply glue only along the bottom and side edges, then attach bamboo pockets to the inside of the folder. Glue the cover art to the front of the folder. Decorate the border around the game board. Glue an envelope to the back of the folder for panda storage. Option: Reproduce, cut out, and glue the Pocket Game Playing Directions to the back of the folder. Note: Make additional "Peek-a-boo Pandas" pocket folders with different combinations of bamboo baskets.

Directions

Reproduce, cut out, and glue the direction strip to the game board.

Place each panda in the bamboo basket with the matching number of dots.

Peek-a-boo Pandas Cover

Counting Number Sets 1-10

Pocket Game

Peek-a-boo Pandas

Pocket Game

Counting Number Sets 1-10

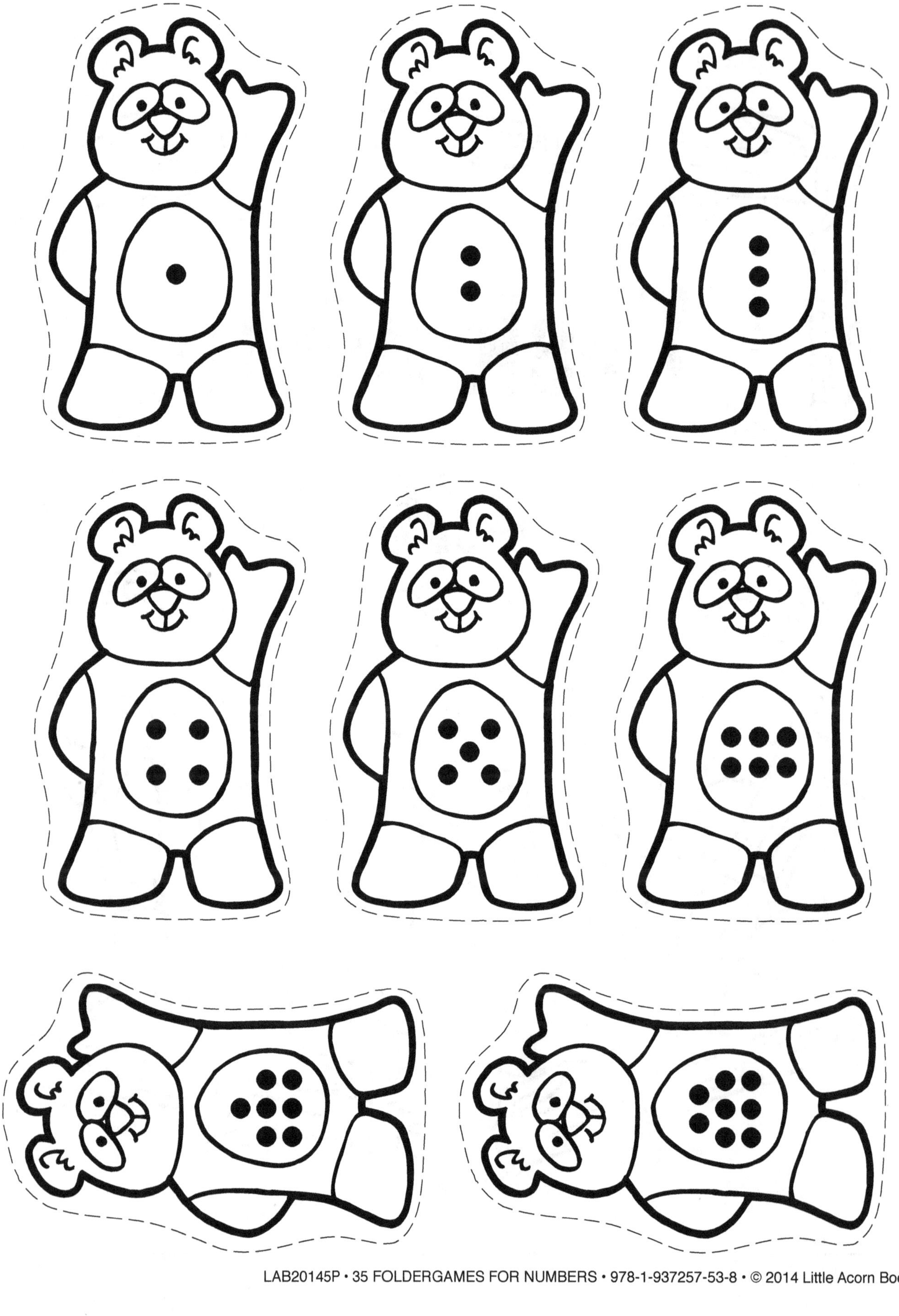

LAB20145P • 35 FOLDERGAMES FOR NUMBERS • 978-1-937257-53-8 •

Peek-a-boo Pandas
Counting Number Sets 1-10

Pocket Game

Include the bear on the right for advanced skills practice.

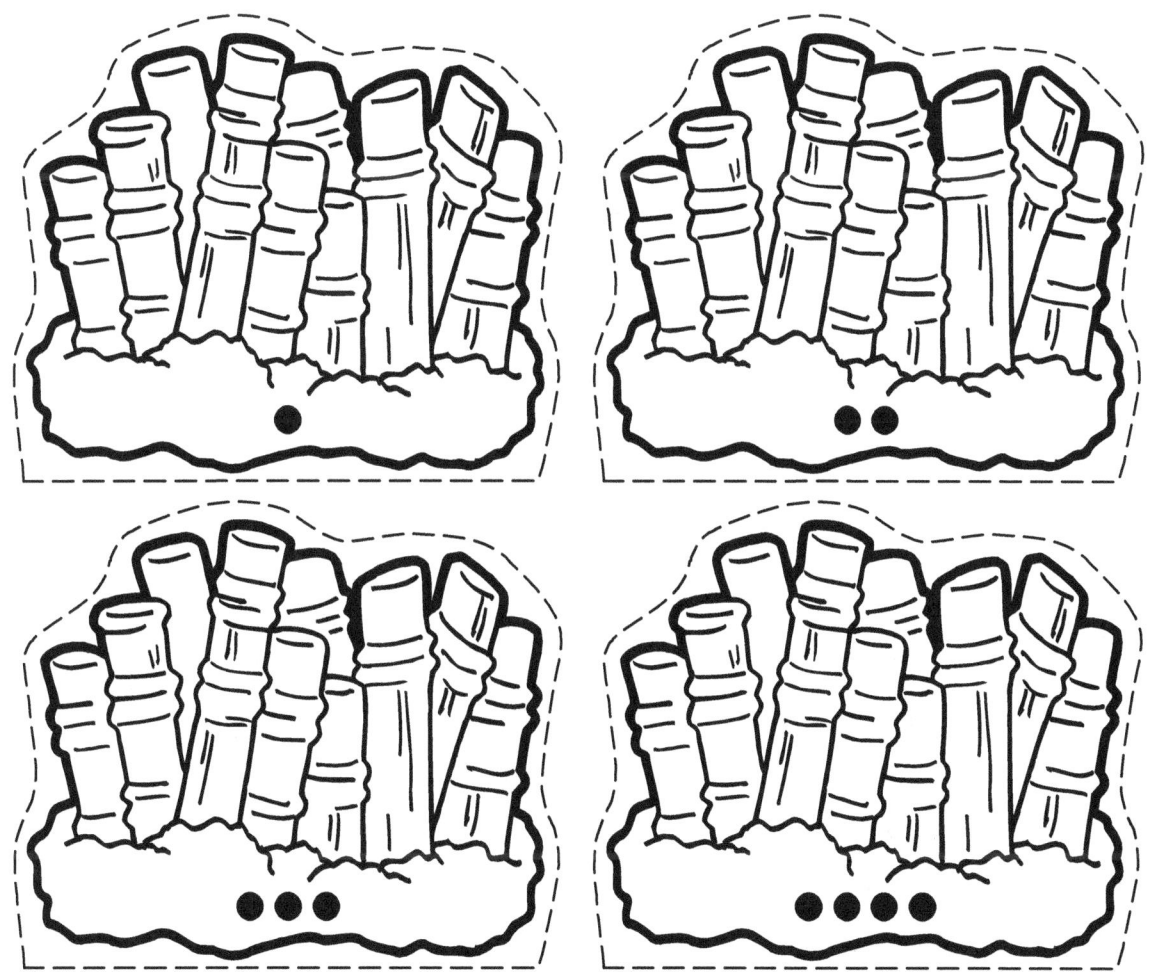

Apply glue only along the bottom and side edges of each bamboo pocket.
Attach each bamboo pocket to the inside of a folder.

Peek-a-boo Pandas
Counting Number Sets 1-10

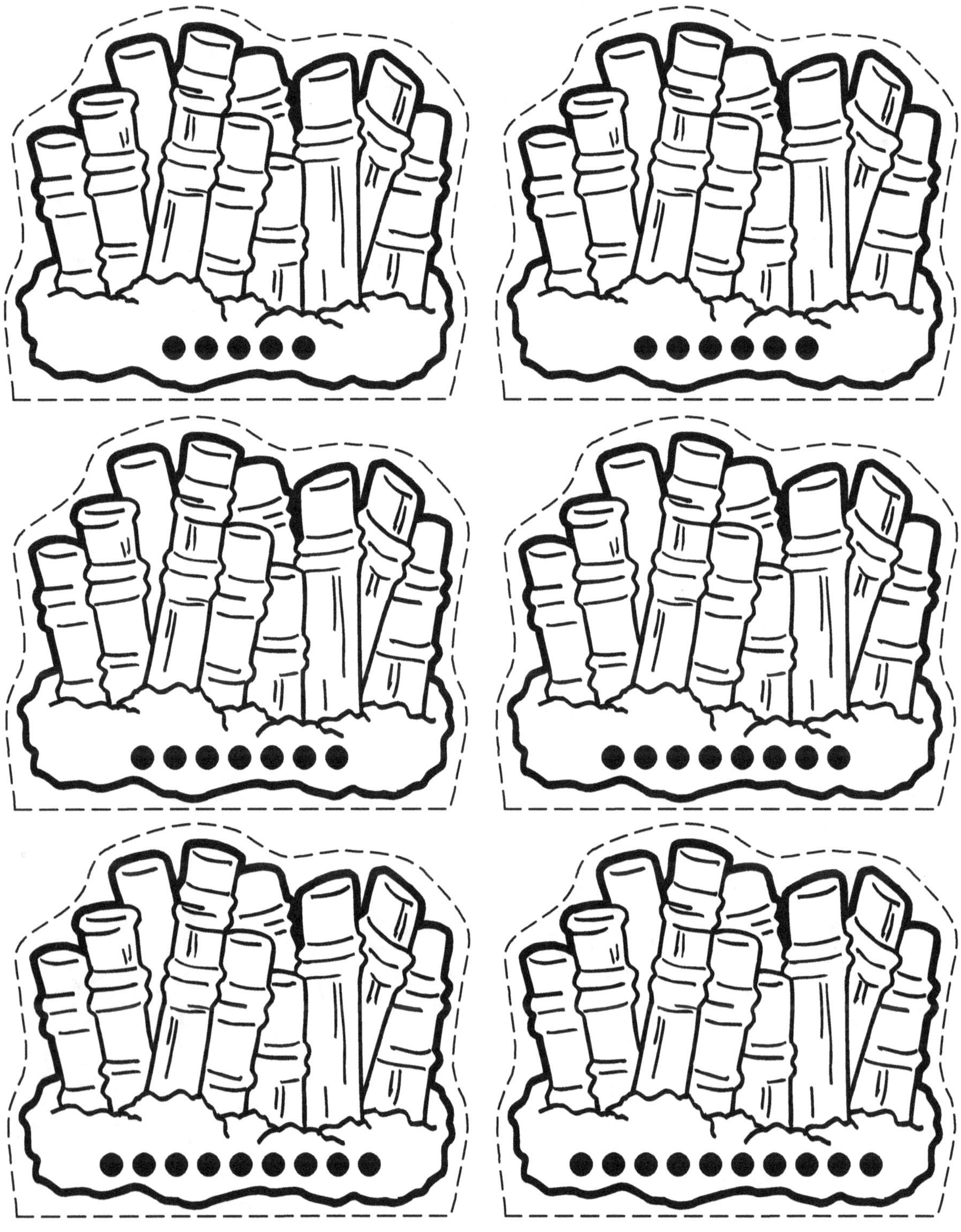

Pocket Game

Apply glue only along the bottom and side edges of each bamboo pocket.
Attach each bamboo pocket to the inside of a folder.

LAB20145P • 35 FOLDERGAMES FOR NUMBERS • 978-1-937257-53-8 • © 2014 Little Acorn Books™

Busy Bees

Matching Number Sets and Numerals 1-10

Fill-ins

Assembly

Reproduce, cut out, and glue a set of oak tag handles (p. 192) to a standard-size file folder. Option: Decorate the handles with yellow and black stripes. Reproduce, color, and cut out the "Busy Bees" patterns. Glue the fill-in boards to the inside of the folder. Glue the cover art to the front of the folder. Decorate the border around the fill-in boards, then laminate. Glue an envelope to the back of the folder for wipe-off marker or crayon storage. Options: Reproduce, cut out, and glue the Fill-ins Playing Directions to the back of the folder. Make different folders to practice different numbers.

Busy Bees Cover
Matching Number Sets and Numerals 1-10

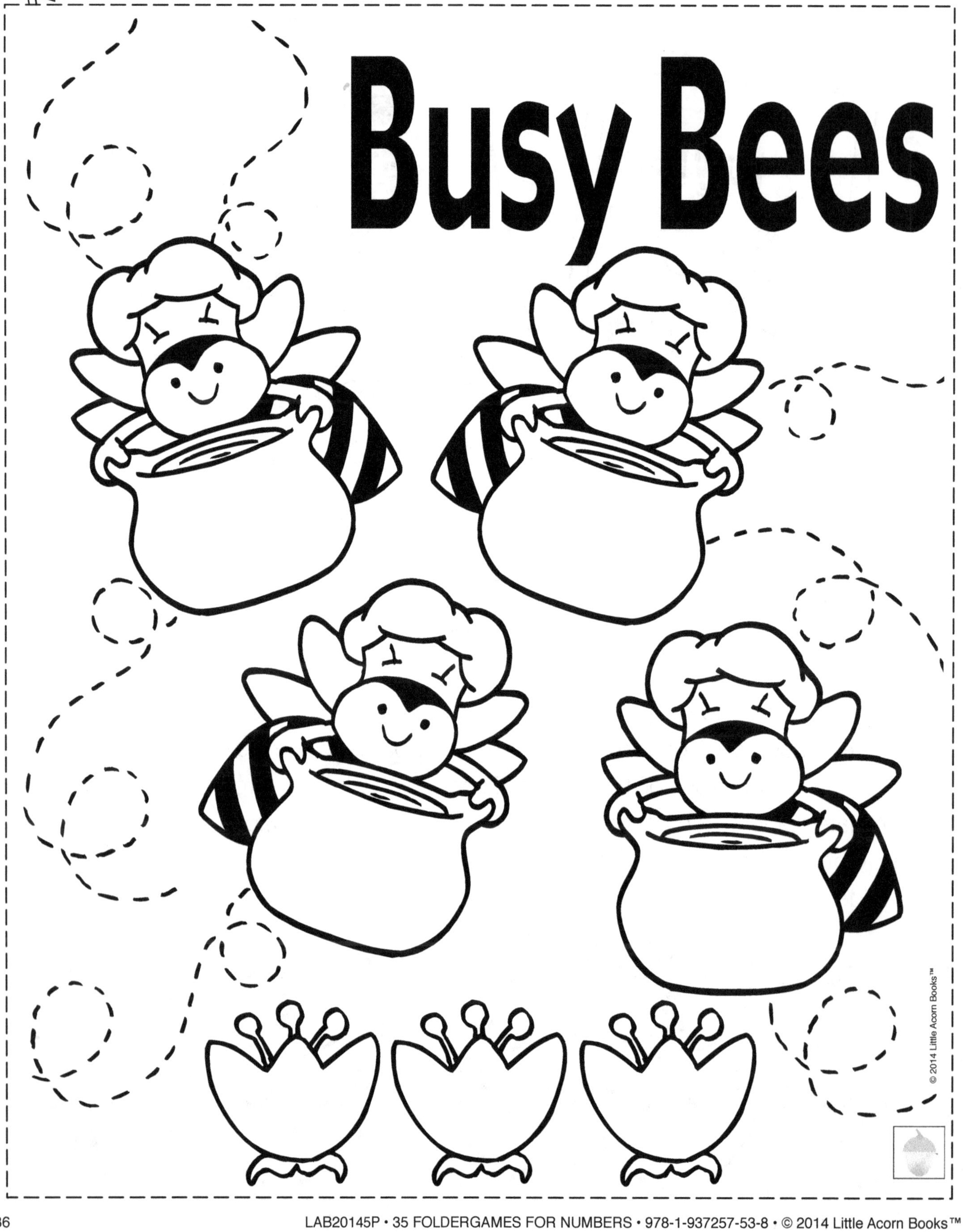

Fill-ins

Busy Bees

Busy Bees
Matching Number Sets and Numerals 1-10

Fill-ins

Color, cut out, and glue two to four placement templates to the inside of a folder. Glue programmed bees and blank flower forms on each template for players to write in missing numerals.

Directions
Reproduce, cut out, and glue the direction strip to the game board.

Busy Bees

Count the dots on each pot. Write the matching numeral on the flower.

Busy Bees
Matching Number Sets and Numerals 1-10

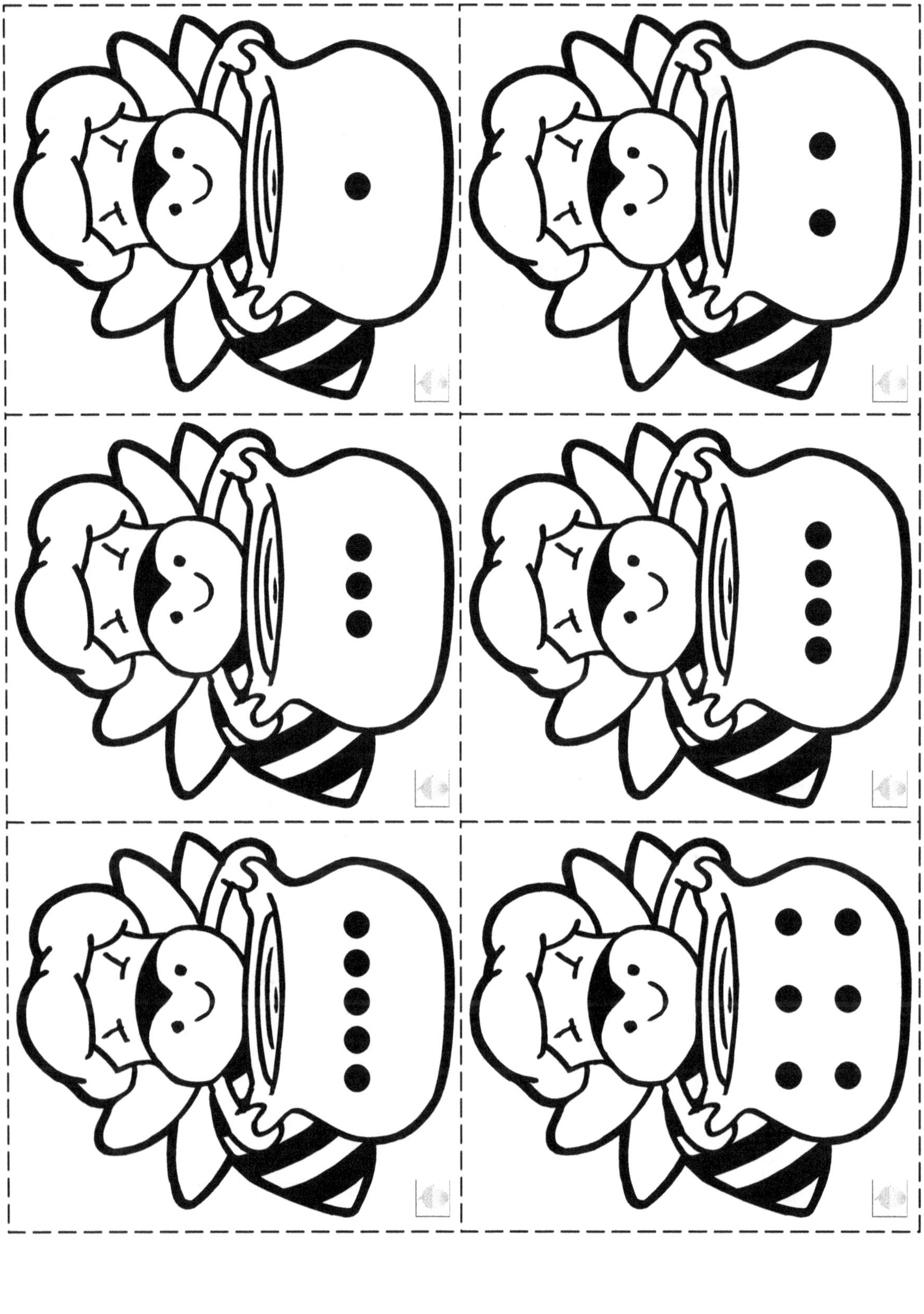

Fill-ins

Busy Bees
Matching Number Sets and Numerals 1-10

Fill-ins

Option: Reproduce, color, cut out, and glue blank bee patterns and programmed flower forms (p. 35) for players to practice drawing the matching number of dots on each bee.

Firefly Fiesta

Matching Number Sets and Numerals 1-10

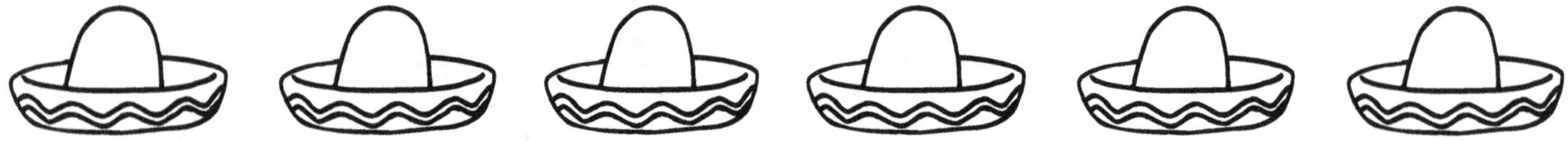

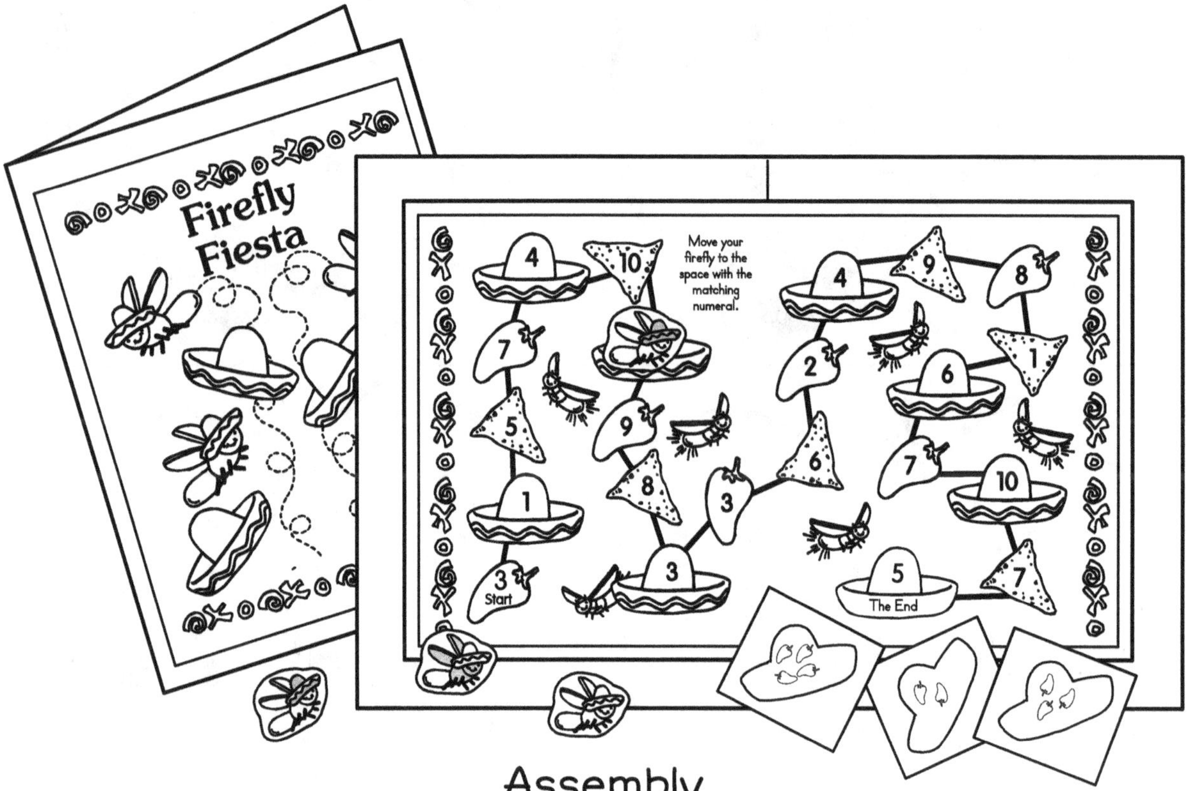

Trail Game

Assembly

Reproduce, cut out, and glue a set of oak tag handles (p. 192) to a standard-size file folder. Option: Decorate each handle with a copy of the hats shown on this page. Reproduce, color, and cut out the "Firefly Fiesta" patterns. Matching in the center, glue the game board patterns to the inside of the folder. Glue the cover art to the front of the folder. Decorate the border around the game board, then laminate. Reproduce, color, laminate, then cut out the pawns and two sets of game cards. Glue an envelope to the back of the folder for pawn and game card storage. Option: Reproduce, cut out, and glue the Trail Game Playing Directions to the back of the folder.

Pawns

Reproduce, color, and cut out.

Firefly Fiesta Cover
Matching Number Sets and Numerals 1-10

Trail Game

Firefly Fiesta

Firefly Fiesta
Matching Number Sets and Numerals 1-10

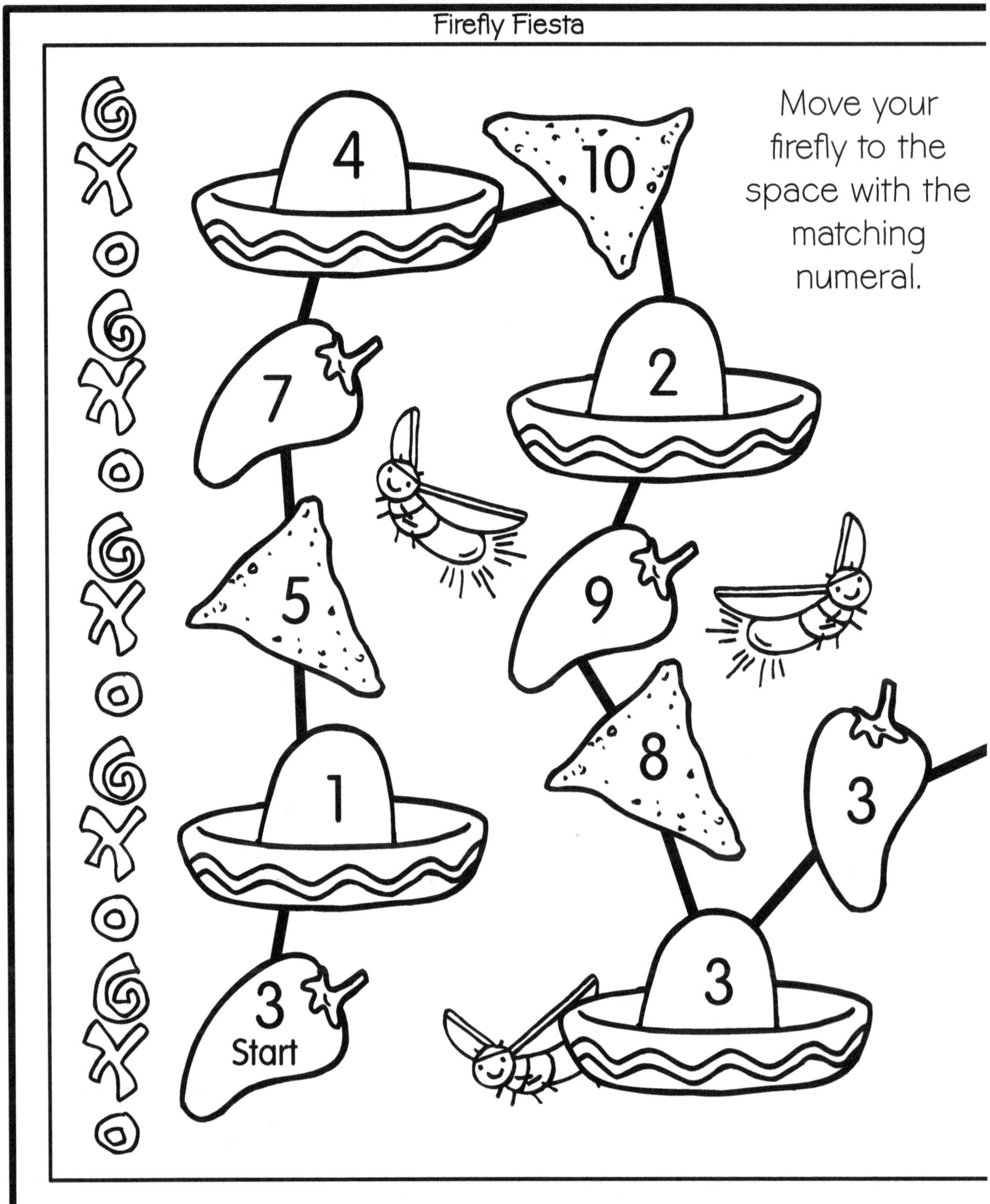

Trail Game

Firefly Fiesta

Move your firefly to the space with the matching numeral.

Firefly Fiesta
Matching Number Sets and Numerals 1-10

Trail Game

Firefly Fiesta
Matching Number Sets and Numerals 1-10

Trail Game

Reproduce, color, and cut apart two sets of cards.

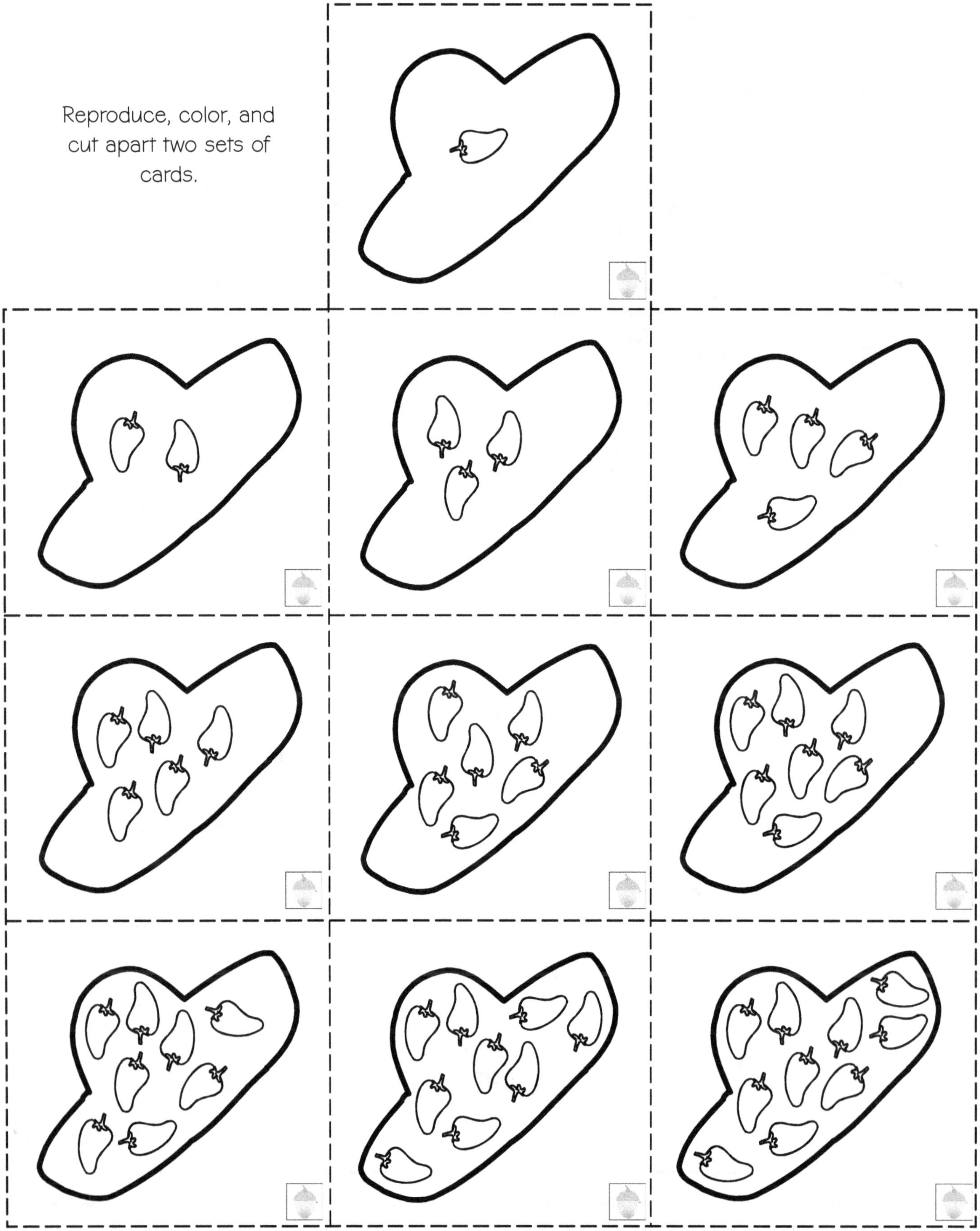

Butterfly Wings

Matching Number Sets and Numerals 1-10

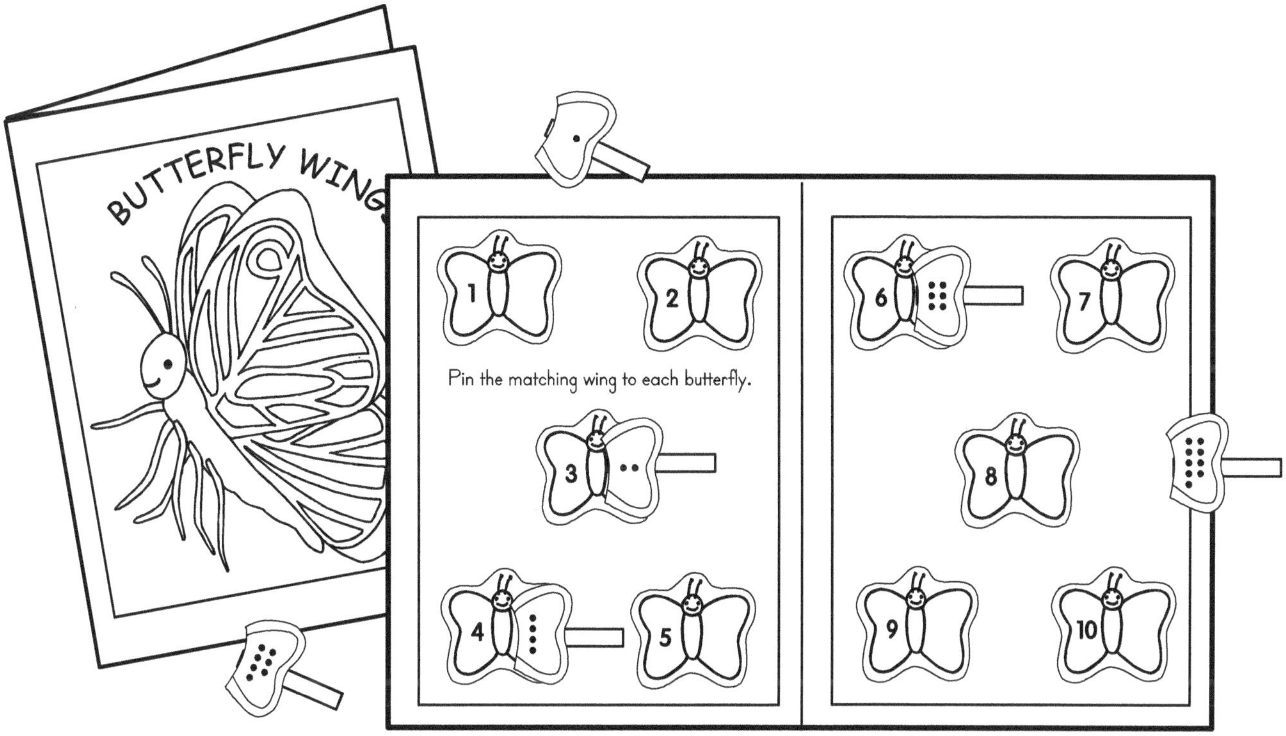

Clothespin Match-ups

Assembly

Reproduce, cut out, and glue a set of oak tag handles (p. 192) to a standard-size file folder. Option: Decorate the handles with copies of the butterflies shown on this page. Reproduce, color, and cut out the "Butterfly Wings" patterns. Glue the wings on clothespins. Arrange and glue the assembled butterflies to the inside of the folder. Glue the cover to the front of the folder. Decorate the border around the butterflies. Staple a resealable plastic bag to the back of the folder for clothespin storage. Option: Reproduce, cut out, and glue the Clothespin Match-ups Playing Directions to the back of the folder.

Butterfly Wings Cover
Matching Number Sets and Numerals 1-10

Clothespin Match-ups

BUTTERFLY WINGS

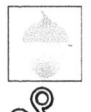

Butterfly Wings
Matching Number Sets and Numerals 1-10

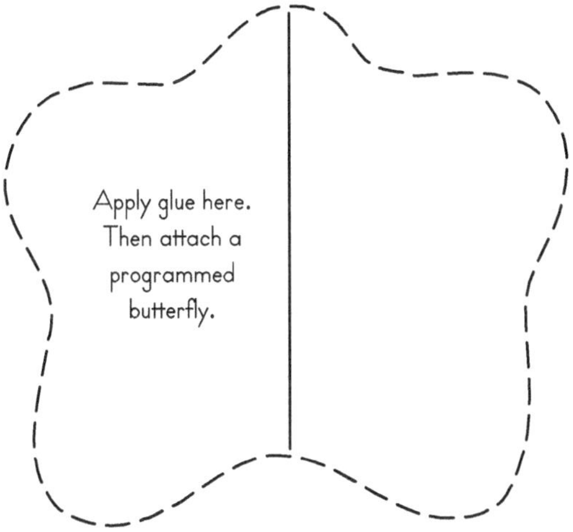

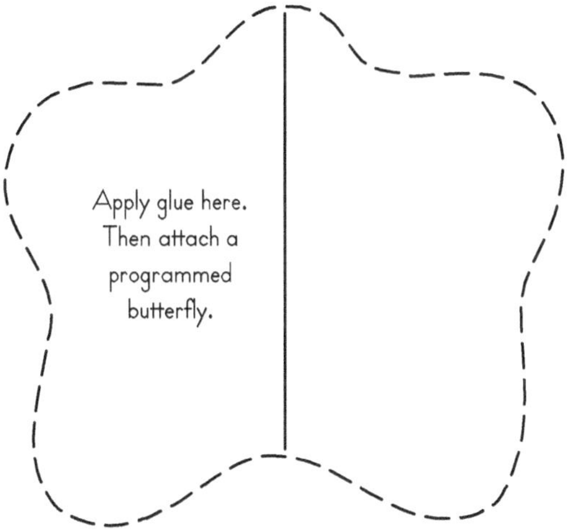

Clothespin Match-ups

Apply glue here. Then attach a programmed butterfly.

Apply glue here. Then attach a programmed butterfly.

Pin the matching wing to each butterfly.

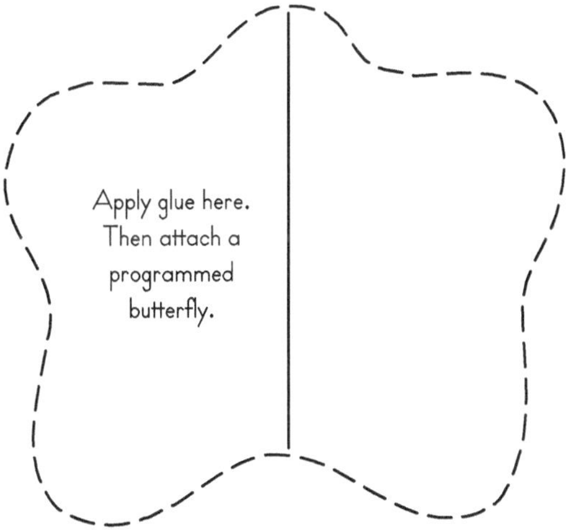

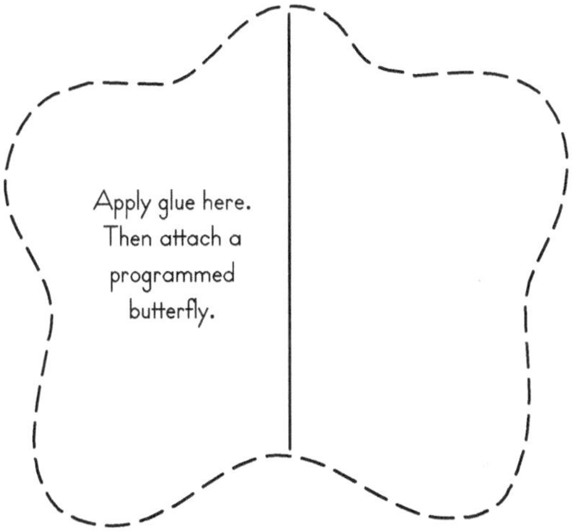

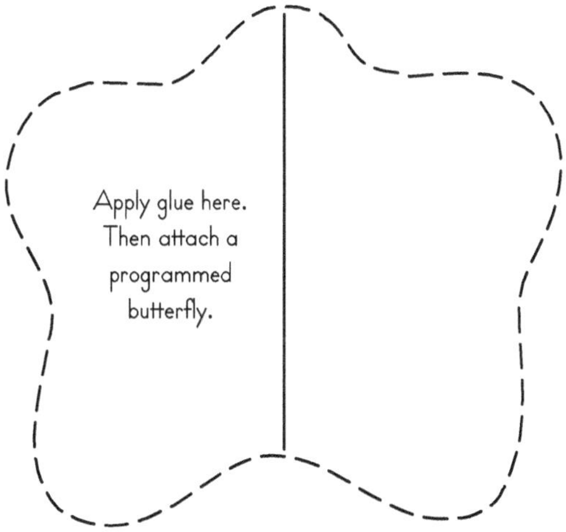

Apply glue here. Then attach a programmed butterfly.

Apply glue here. Then attach a programmed butterfly.

Apply glue here. Then attach a programmed butterfly.

Butterfly Wings
Matching Number Sets and Numerals 1-10

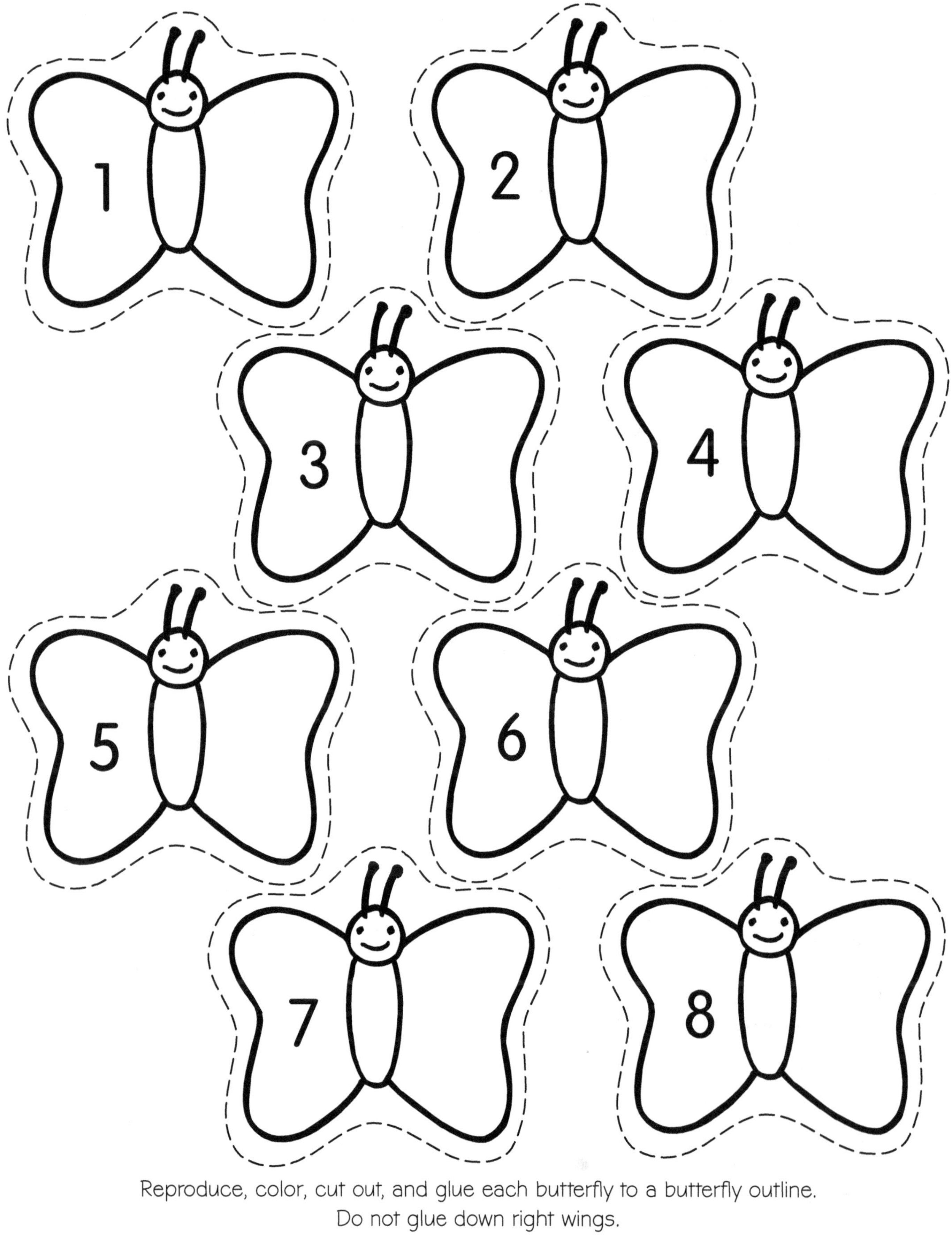

Clothespin Match-ups

1 2
3 4
5 6
7 8

Reproduce, color, cut out, and glue each butterfly to a butterfly outline.
Do not glue down right wings.

Butterfly Wings
Matching Number Sets and Numerals 1-10

Clothespin Match-ups

Reproduce, color, cut out, and glue each butterfly to a butterfly outline.

Reproduce, color, cut out, and glue each wing on a clothespin.

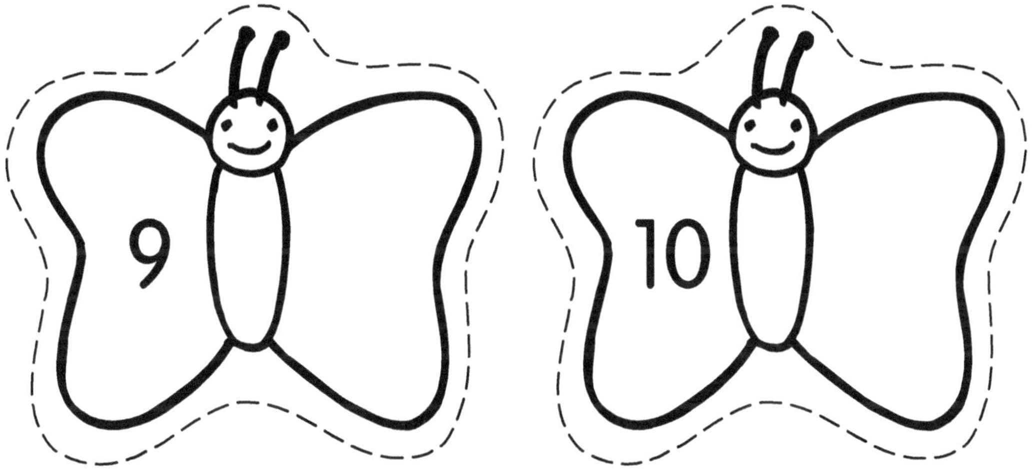

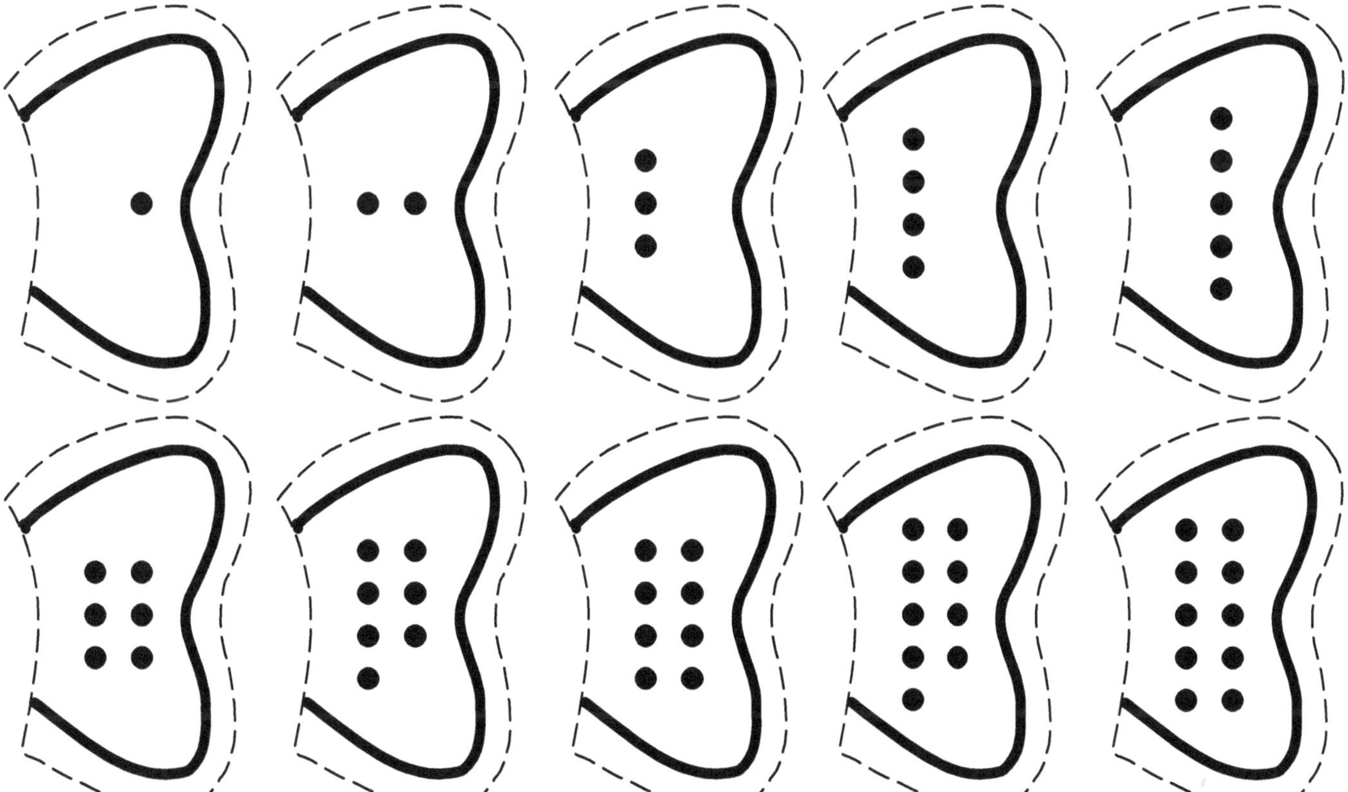

Reproduce, color, cut out, and glue these wings on clothespins for advanced skills practice.

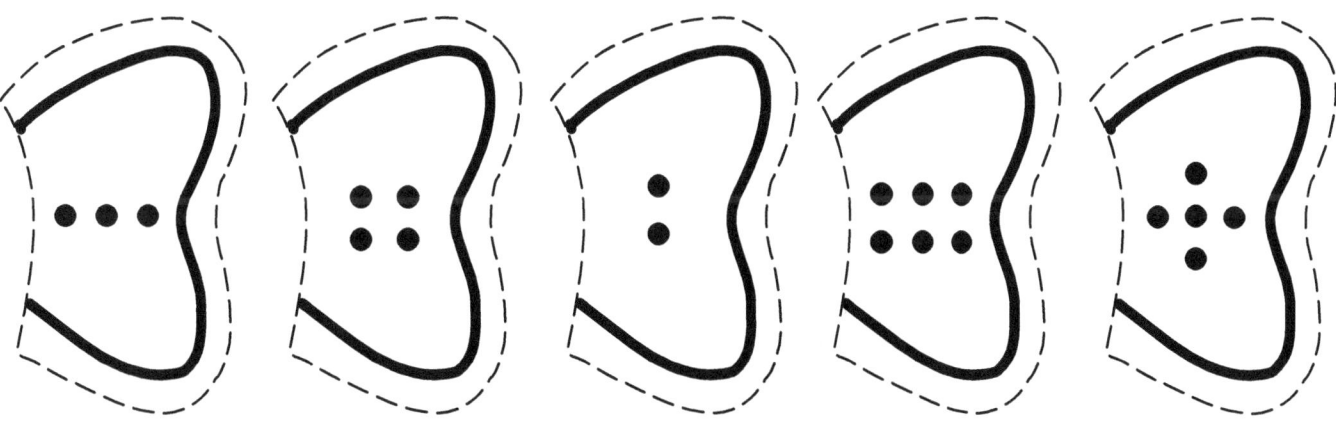

A Spider's Adventure

Matching Number Sets and Numerals 1-10

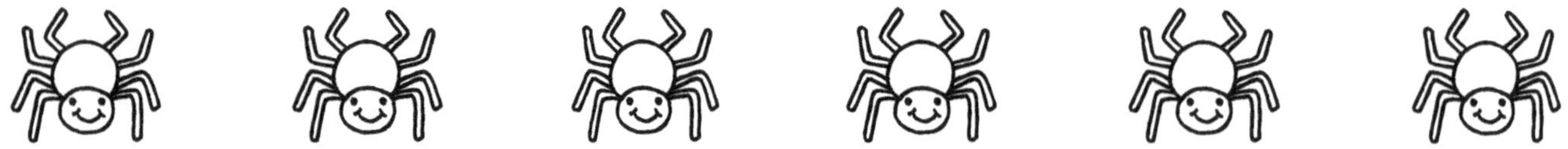

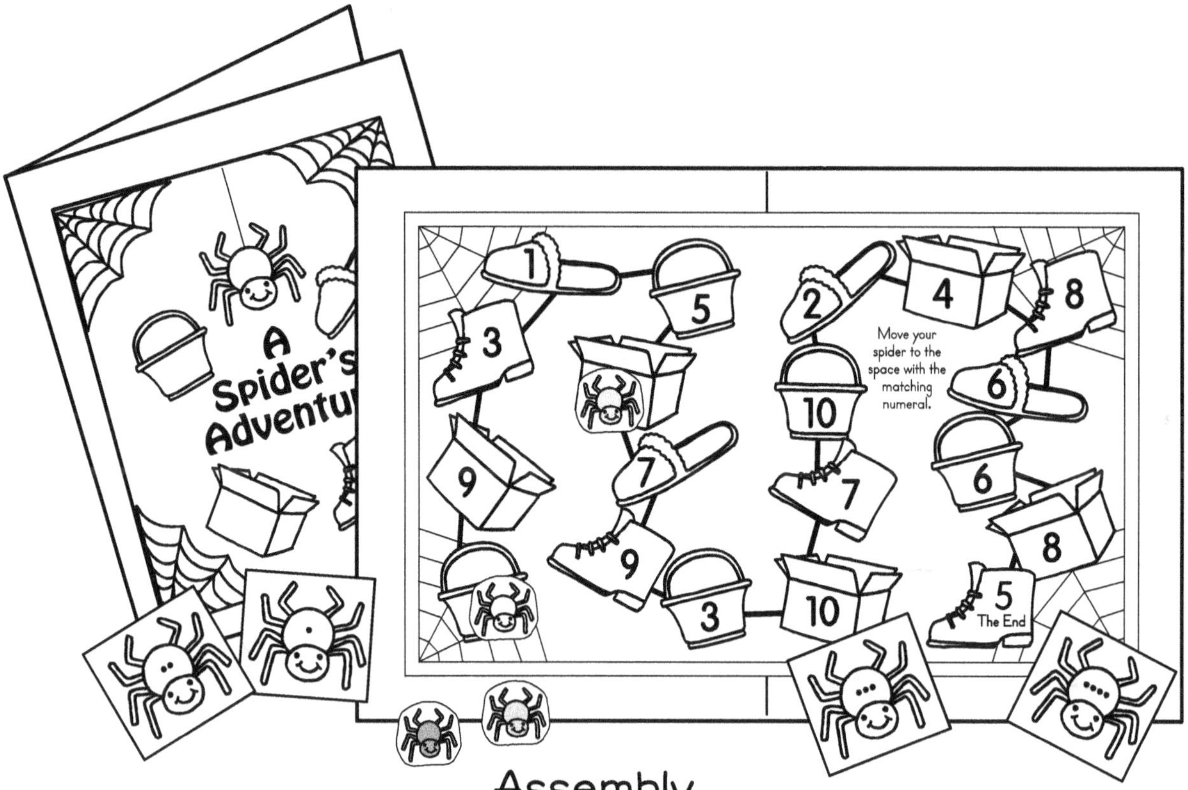

Trail Game

Assembly

Reproduce, cut out, and glue a set of oak tag handles (p. 192) to a standard-size file folder. Option: Decorate each handle with a copy of the spiders on this page. Reproduce, color, and cut out the "A Spider's Adventure" patterns. Matching in the center, glue the game board patterns to the inside of the folder. Glue the cover art to the front of the folder. Decorate the border around the game board, then laminate. Reproduce, color, laminate, then cut out the pawns and two sets of game cards. Glue an envelope to the back of the folder for pawn and game card storage. Option: Reproduce, cut out, and glue the Trail Game Playing Directions to the back of the folder.

Pawns

Reproduce, color each spider a different color, and cut out.

A Spider's Adventure Cover

Matching Number Sets and Numerals 1-10

Trail Game

A Spider's Adventure

A Spider's Adventure
Matching Number Sets and Numerals 1-10

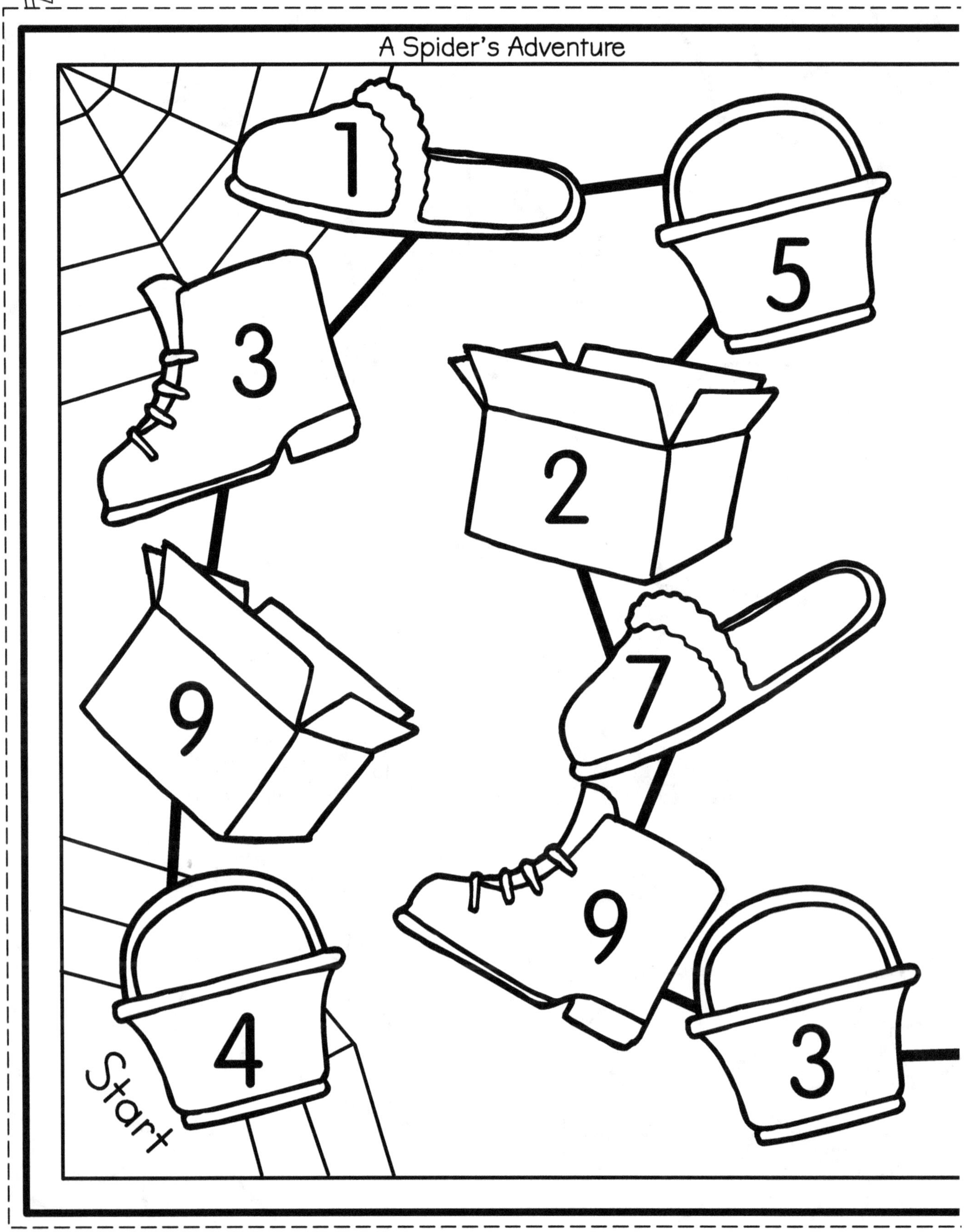

Trail Game

A Spider's Adventure

Matching Number Sets and Numerals 1-10

Trail Game

A Spider's Adventure

Move your spider to the space with the matching numeral.

2, 4, 8, 10, 6, 7, 6, 10, 8, 5 — The End

© 2014 Little Acorn Books™

A Spider's Adventure
Matching Number Sets and Numerals 1-10

Trail Game

Reproduce, color, and cut apart two sets of cards.

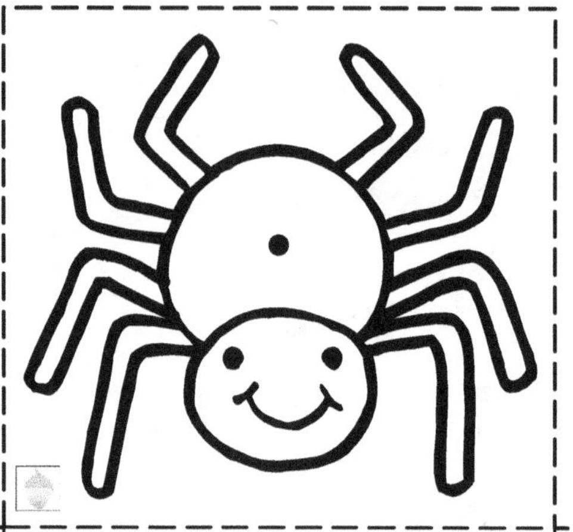

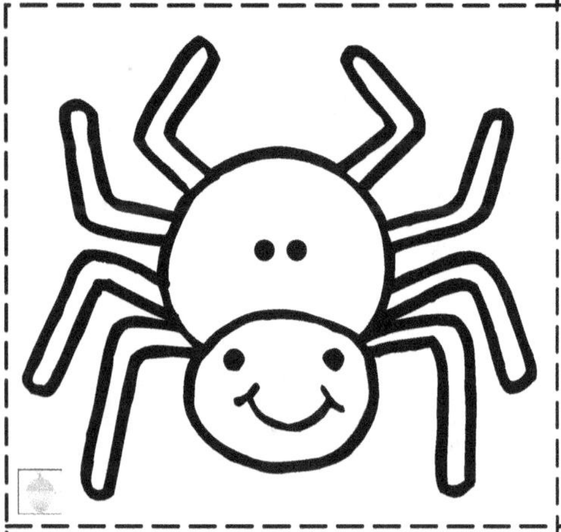

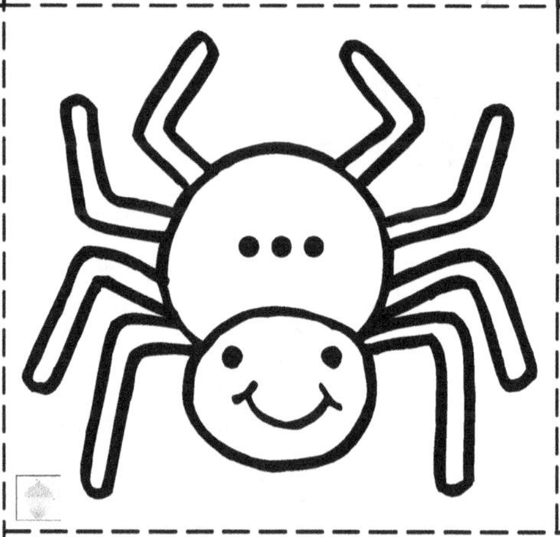

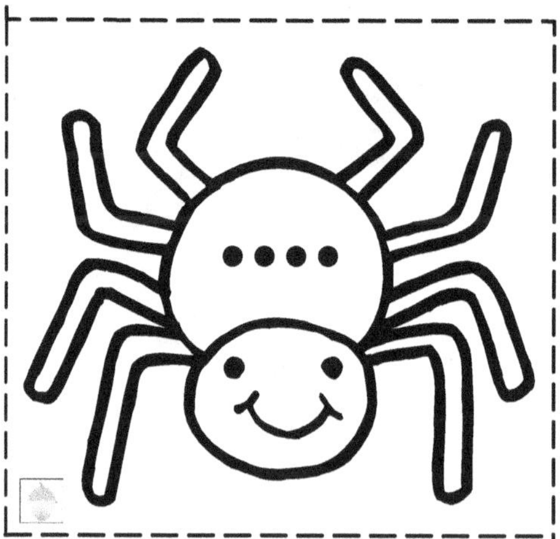

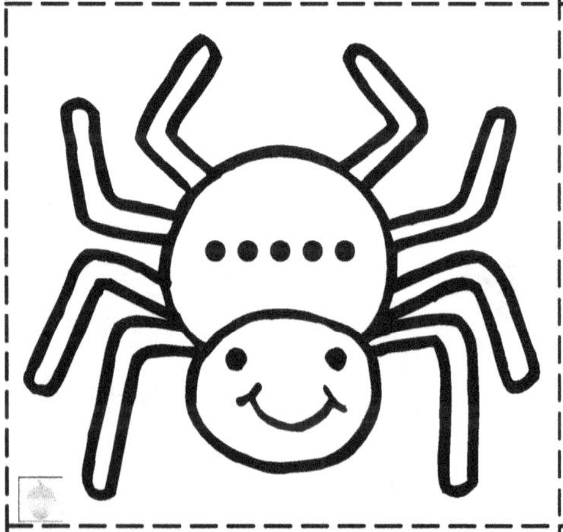

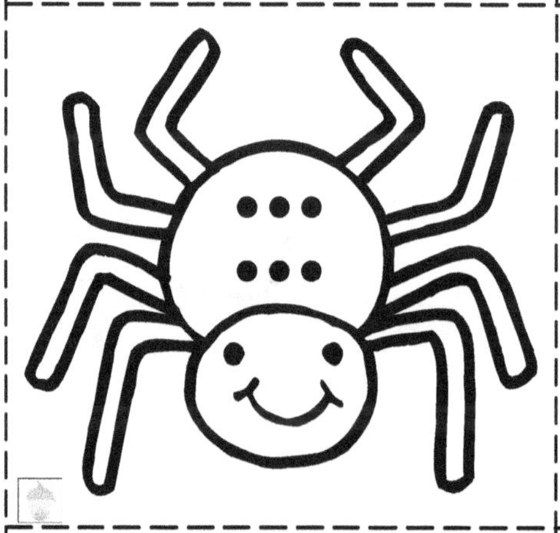

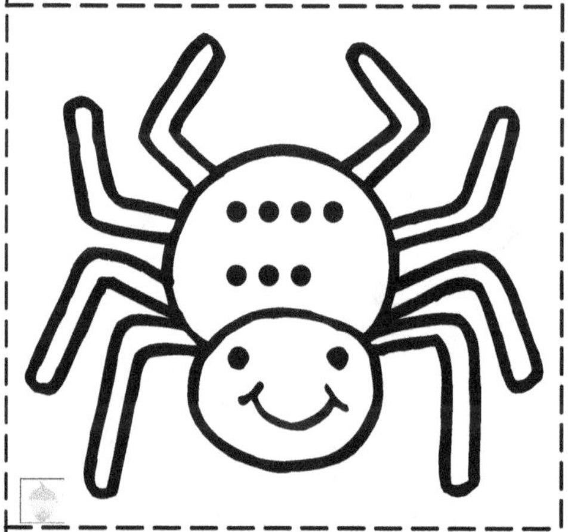

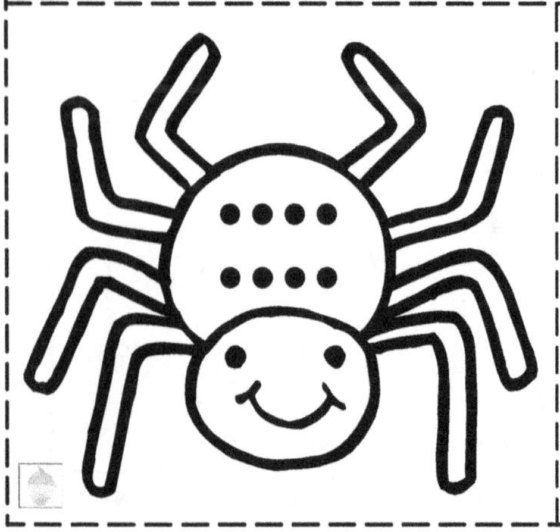

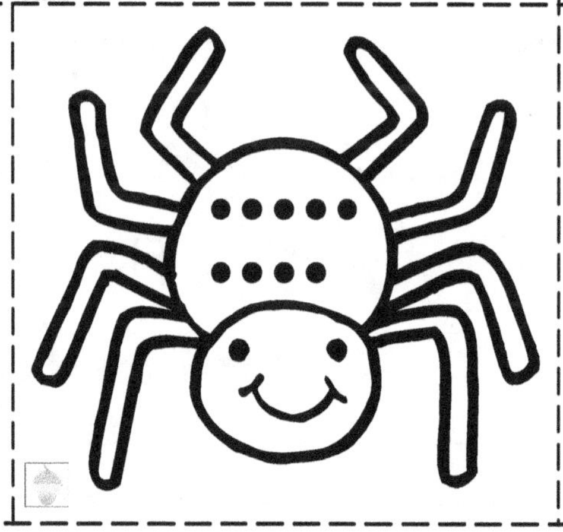

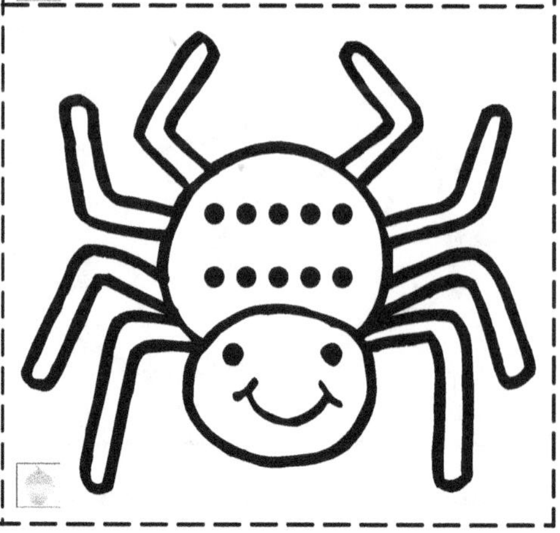

Peek-a-boo Monkeys
Sequencing Number Sets and Numerals 1-10

Pocket Game

Assembly

Reproduce, cut out, and glue a set of oak tag handles (p. 192) to a standard-size file folder. Option: Decorate each handle with a copy of the monkeys on this page. Reproduce, color, and cut out the "Peek-a-boo Monkeys" patterns. Apply glue only along the bottom and side edges, then attach barrels to the inside of the folder. Glue the cover art to the front of the folder. Decorate the border around the game board. Glue an envelope to the back of the folder for monkey storage. Option: Reproduce, cut out, and glue the Pocket Game Playing Directions to the back of the folder. Note: Make additional "Peek-a-boo Monkeys" pocket folders with different combinations of baskets.

Directions

Reproduce, cut out, and glue the direction strip to the game board.

Place each monkey in the matching barrel.

Peek-a-boo Monkeys Cover

Sequencing Number Sets and Numerals 1-10

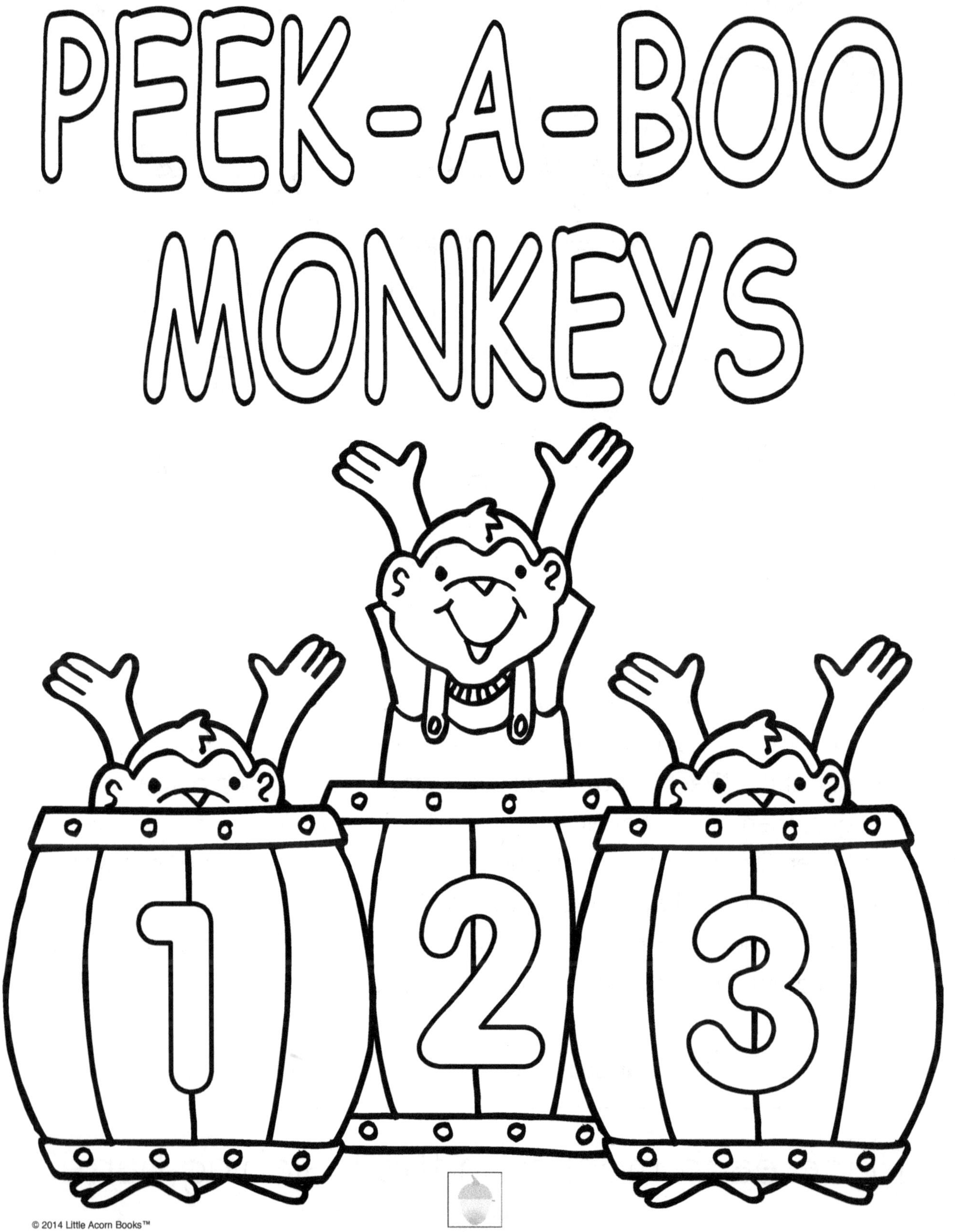

Pocket Game

Peek-a-boo Monkeys
Sequencing Number Sets and Numerals 1-10

Pocket Game

Reproduce, color, and cut out the monkeys below for advanced skills practice.

Peek-a-boo Monkeys
Sequencing Number Sets and Numerals 1-10

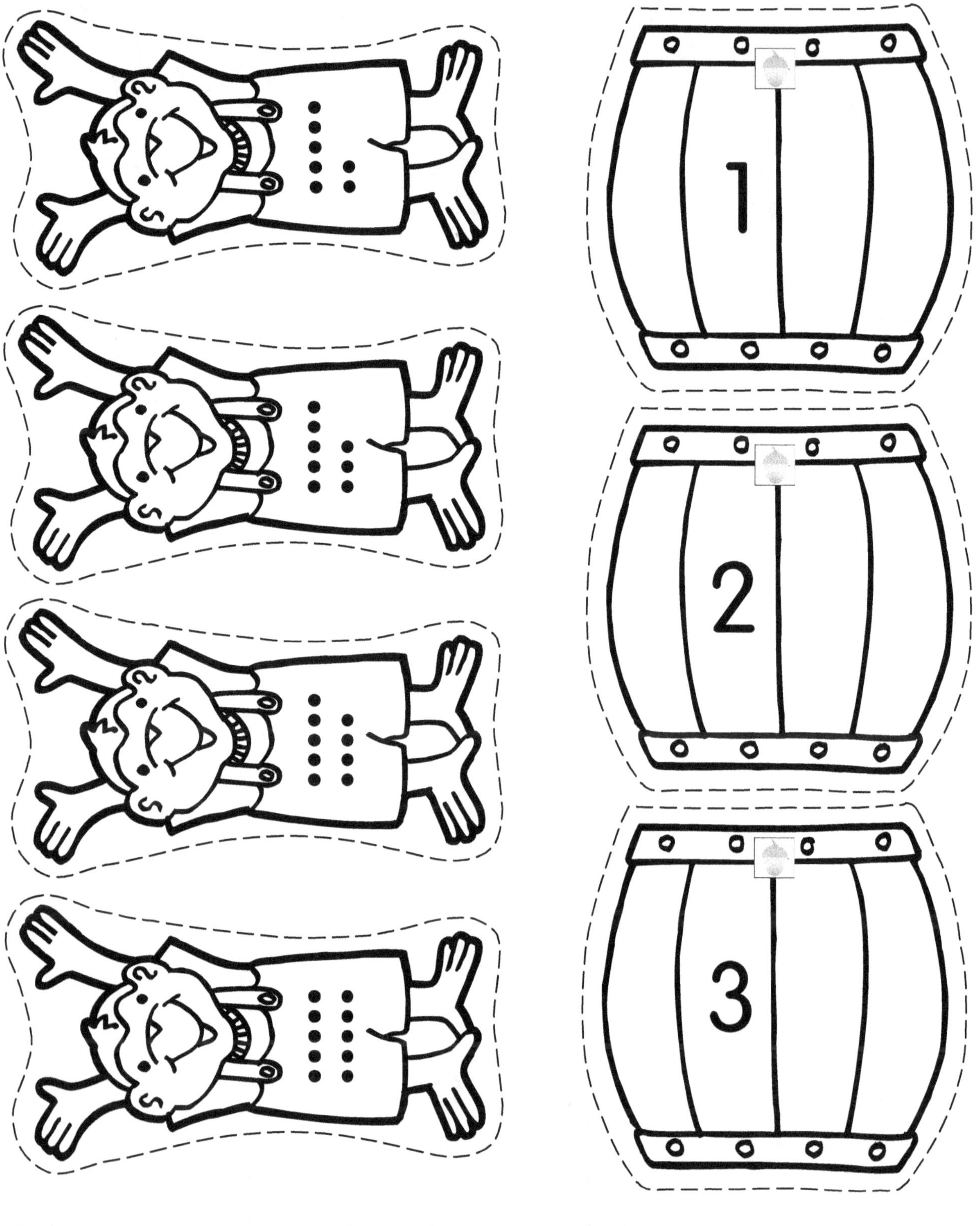

Pocket Game

Reproduce, color, and cut out the monkeys and barrels. Apply glue only along the bottom and side edges of each barrel. Attach each barrel to the inside of a folder.

58   LAB20145P • 35 FOLDERGAMES FOR NUMBERS • 978-1-937257-53-8 • © 2014 Little Acorn Books™

Peek-a-boo Monkeys
Sequencing Number Sets and Numerals 1-10

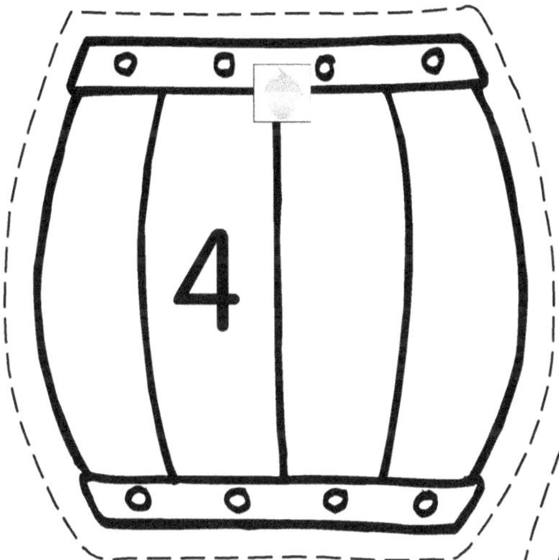

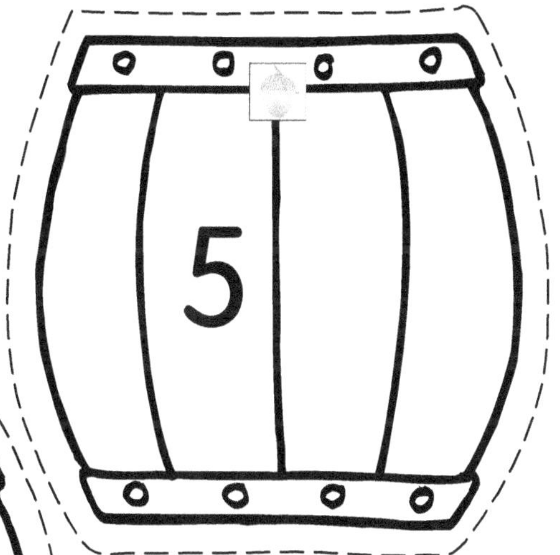

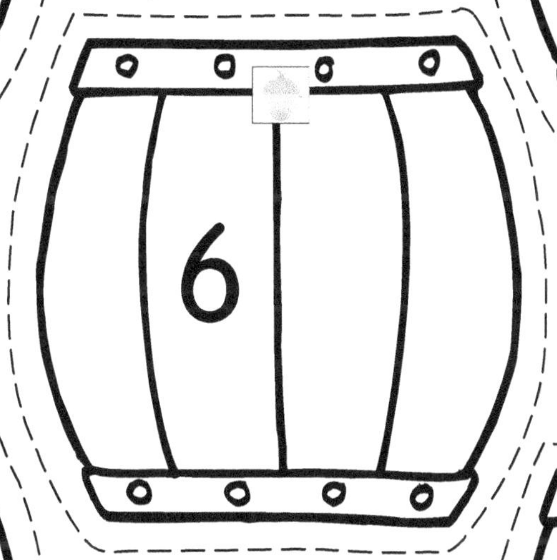

Pocket Game

4

5

6

7

Apply glue only along the bottom and side edges of each barrel. Attach each barrel to the inside of a folder.

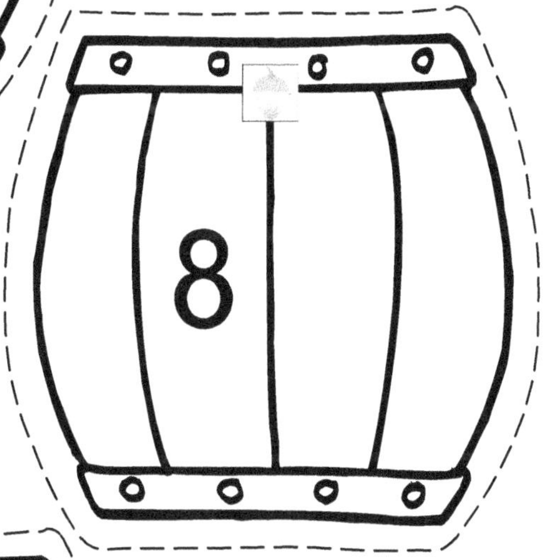

8

9

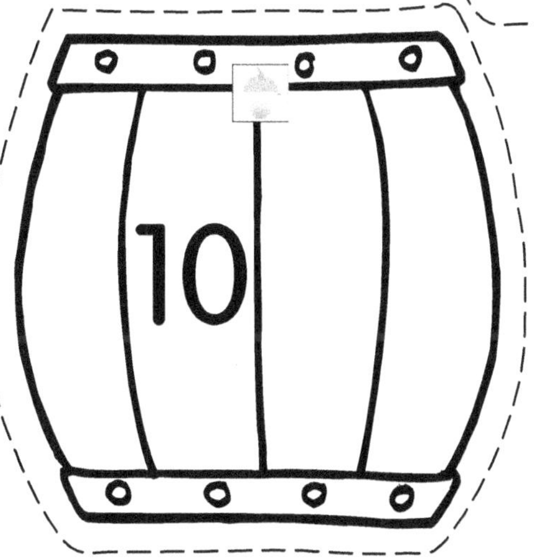

10

LAB20145P • 35 FOLDERGAMES FOR NUMBERS • 978-1-937257-53-8 • © 2014 Little Acorn Books™

59

Polka-Dot Turtles
Sequencing Number Sets 1-10

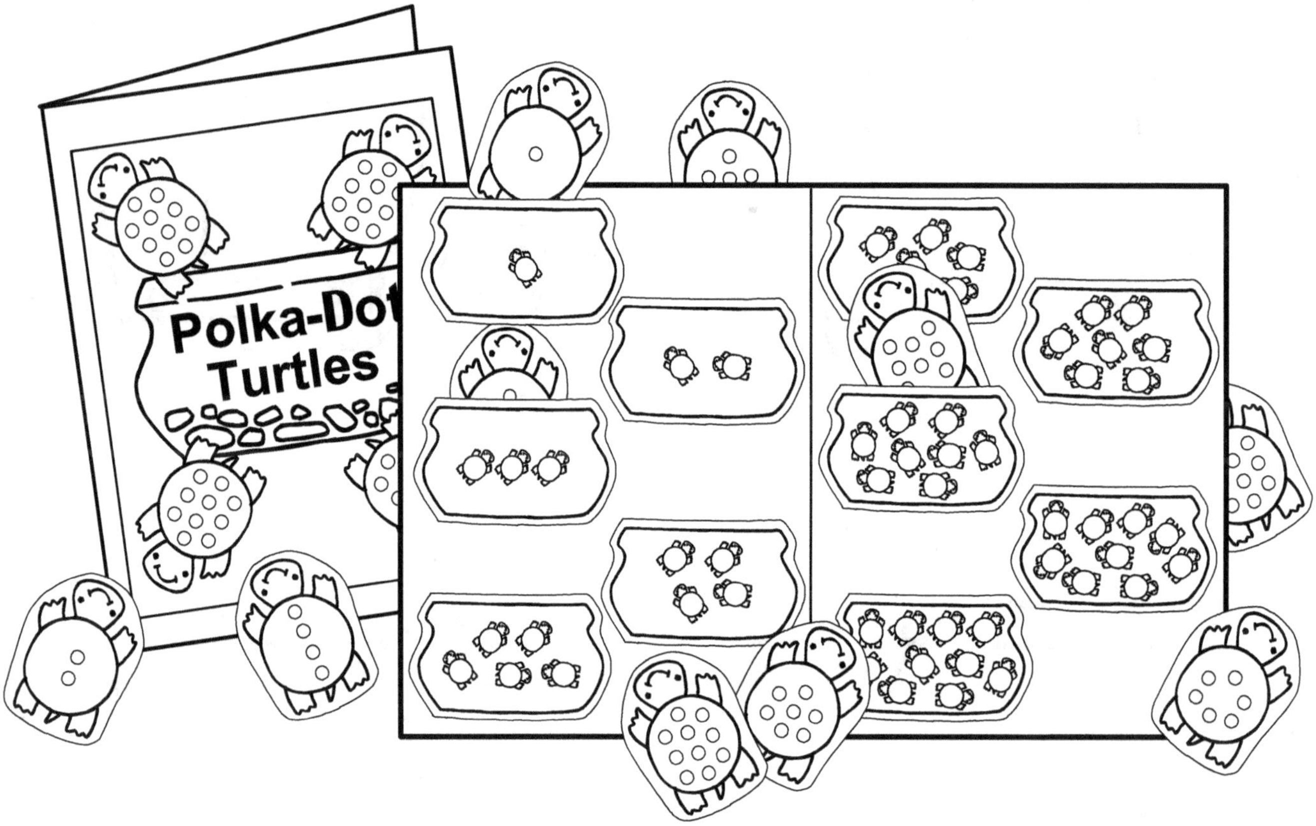

Pocket Game

Assembly

Reproduce, cut out, and glue a set of oak tag handles (p. 192) to a standard-size file folder. Option: Decorate each handle with a copy of the turtles on this page. Reproduce, color, and cut out the "Polka-Dot Turtles" patterns. Apply glue only along the bottom and side edges, then attach bowls to the inside of the folder. Glue the cover art to the front of the folder. Decorate the border around the game board. Glue an envelope to the back of the folder for turtle storage. Option: Reproduce, cut out, and glue the Pocket Game Playing Directions to the back of the folder.

Directions
Reproduce, cut out, and glue the direction strip to the game board.

Place each turtle in the matching bowl.

Polka-Dot Turtles Cover
Sequencing Number Sets 1-10

Pocket Game

Polka-Dot Turtles

© 2014 Little Acorn Books™

LAB20145P • 35 FOLDERGAMES FOR NUMBERS • 978-1-937257-53-8 • © 2014 Little Acorn Books™

Polka-Dot Turtles
Sequencing Number Sets 1-10

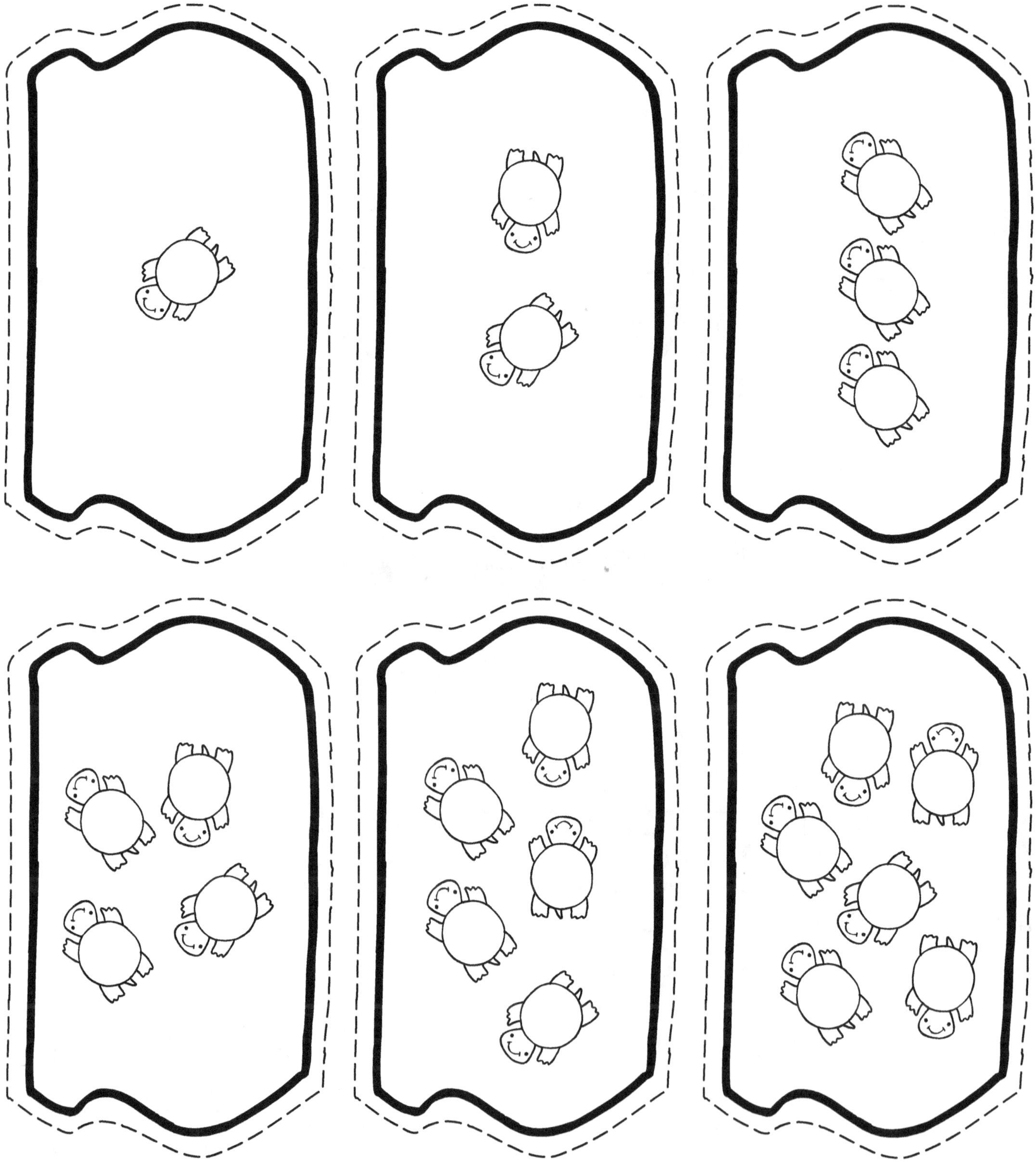

Pocket Game

Reproduce, color, and cut out bowls. Apply glue only along the bottom and side edges of each bowl. Attach each bowl to the inside of a folder.

Polka-Dot Turtles
Sequencing Number Sets 1-10

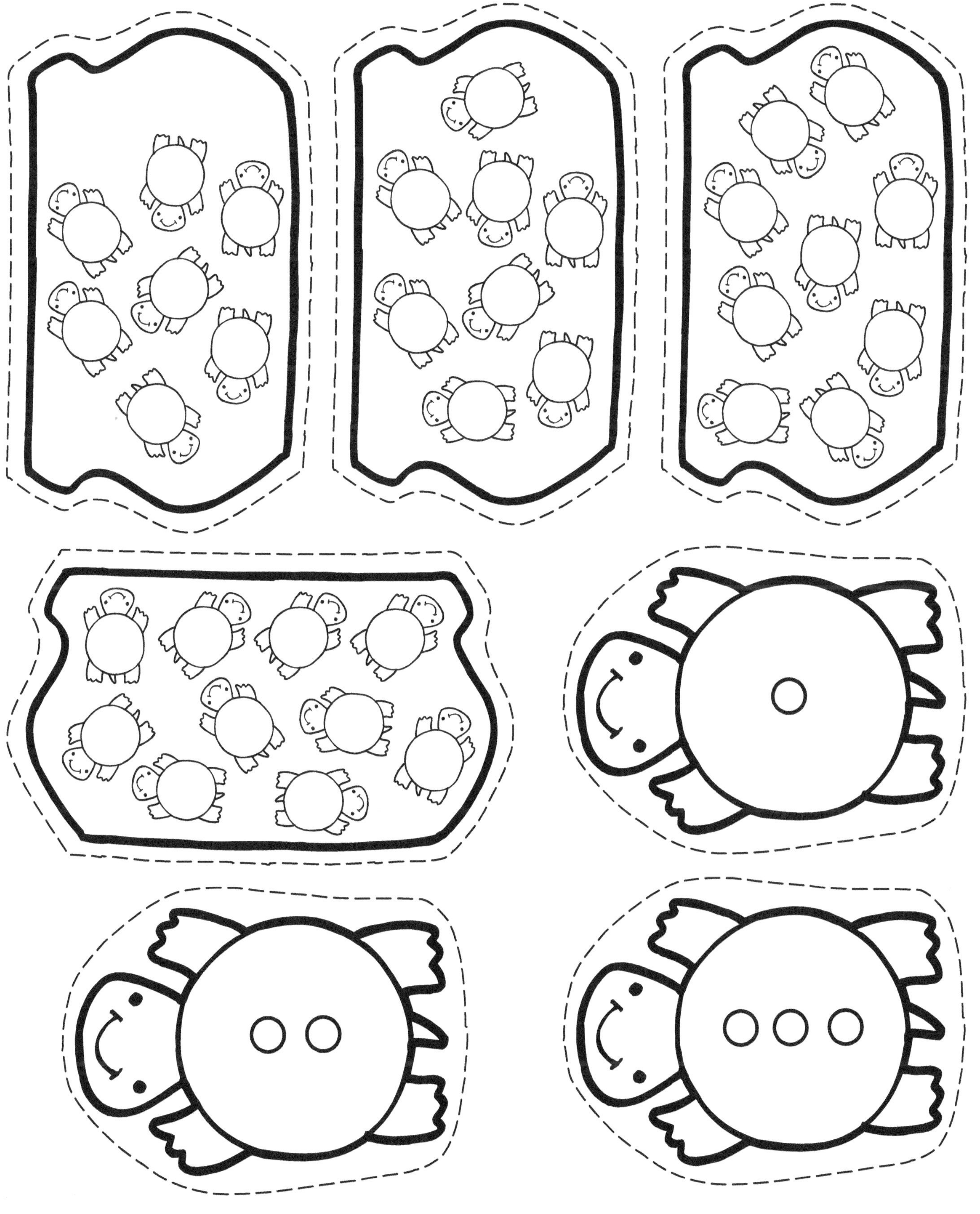

Pocket Game

Polka-Dot Turtles
Sequencing Number Sets 1-10

Pocket Game

Include the turtle on the right for advanced skills practice.

64   LAB20145P • 35 FOLDERGAMES FOR NUMBERS • 978-1-937257-53-8 • © 2014 Little Acorn Books™

Poppin' Popcorn
Sequencing Numerals 1-10

Clothespin Wheel

Assembly

Reproduce, cut out, and glue a set of oak tag handles (p. 192) to a standard-size file folder. Option: Decorate the handles with copies of the popcorn shown on this page. Reproduce, color, and cut out the "Poppin' Popcorn" patterns. Glue the cover art to the front of the folder. Staple a resealable plastic bag to the back of the folder for clothespin storage. Glue the word popped corn kernels on the wheel. Glue the rest of the numeral kernels on clothespins. Meeting along the straight edges, glue the wheel quarters on a large poster board circle. Glue the wheel inside the folder. Option: Reproduce, cut out, and glue the directions (below) to the back of the folder.

Directions

Reproduce, cut out, and glue the direction strip to the back of the folder.

Pin the popcorn around the wheel in numerical order.

Poppin' Popcorn Cover
Sequencing Numerals 1-10

Clothespin Wheel

Poppin' Popcorn

Poppin' Popcorn
Sequencing Numerals 1-10

Clothespin Wheel

Reproduce, color, and cut out four wheel quarters. Meeting along the straight edges, glue the quarters on a large poster board circle. Note: Children do not count the popcorn kernels on the wheel.

Poppin' Popcorn
Sequencing Numerals 1-10

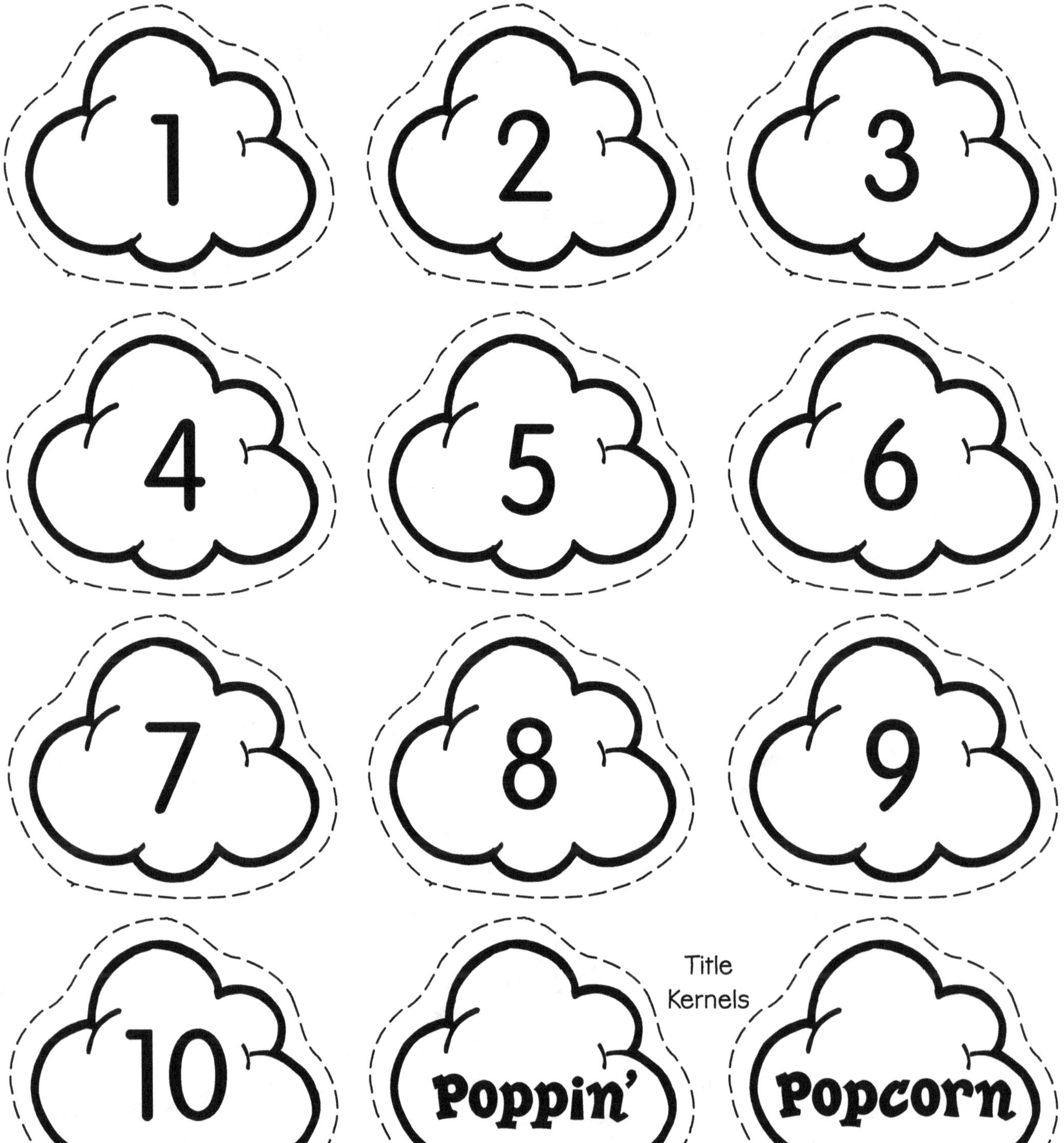

Clothespin Wheel

1   2   3

4   5   6

7   8   9

10   Poppin'   Popcorn

Title Kernels

Reproduce and glue each numbered popped corn kernel on a clothespin.

Marvelous Mouse
Sequencing Numerals 1-10

Dot-to-Dot

Assembly

Reproduce, cut out, and glue a set of oak tag handles (p. 192) to a standard-size file folder. Option: Decorate each handle with a copy of the mice shown on this page. Reproduce, color, and cut out the "Marvelous Mouse" patterns. Matching in the center, glue the mouse to the inside of the folder. Glue the cover art to the front of the folder, then laminate. Laminate, then cut apart, the dot-to-dot pictures. Glue an envelope to the back of the folder to store dot-to-dot pictures, a wipe-off marker or crayon, and soft cloth. Children take out the pictures one at a time to place on the mouse's sign. Option: Reproduce, cut out, and glue the Dot-to-Dot Playing Directions to the back of the folder.

Marvelous Mouse
Sequencing Numerals 1-10

Dot-to-Dot

Reproduce, color, cut out, and glue to the inside of a folder.

Marvelous Mouse
Sequencing Numerals 1-10

Dot-to-Dot

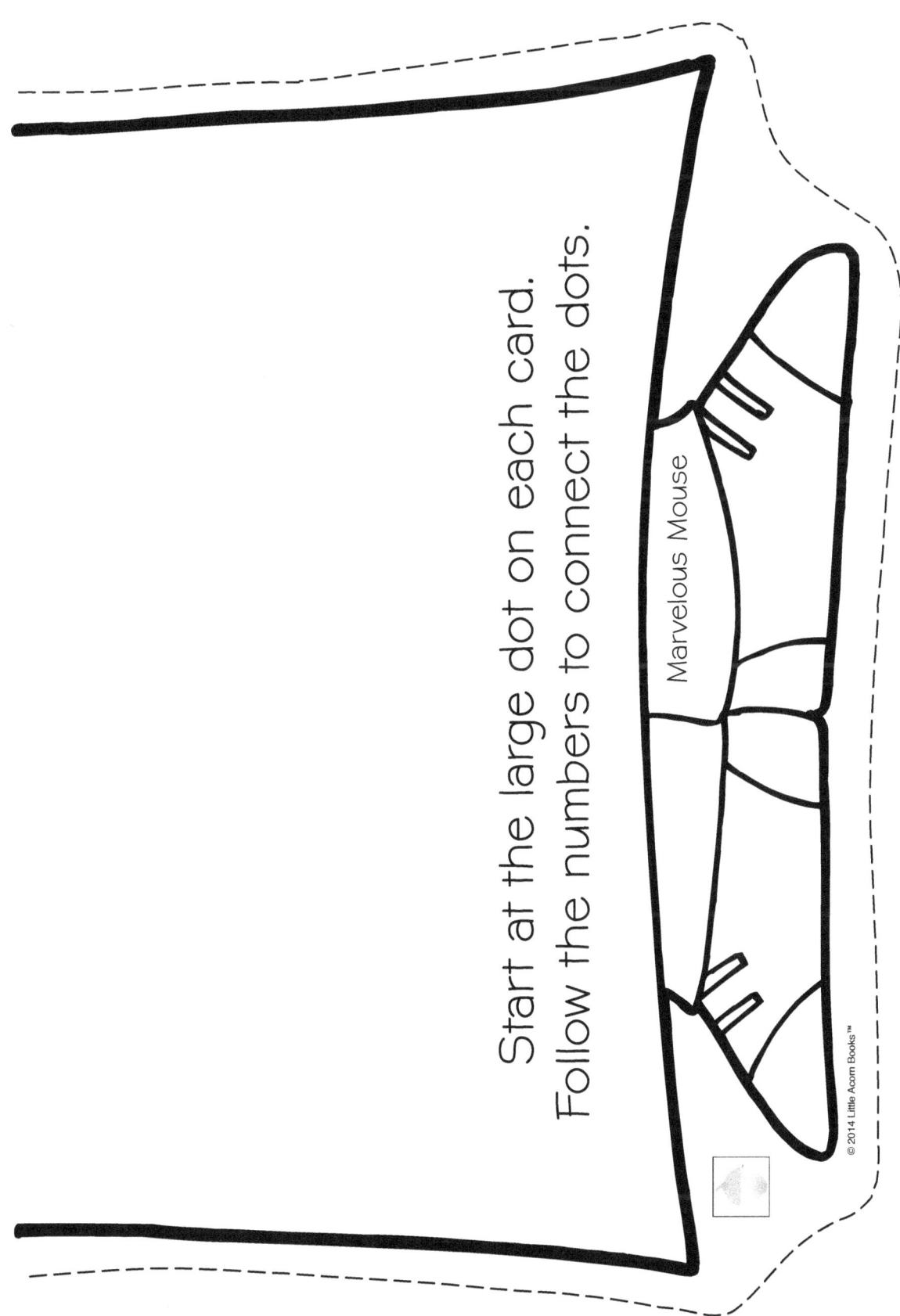

Start at the large dot on each card. Follow the numbers to connect the dots.

Reproduce, color, cut out, and glue to the inside of a folder.

Marvelous Mouse
Sequencing Numerals 1-10
Dot-to-Dot

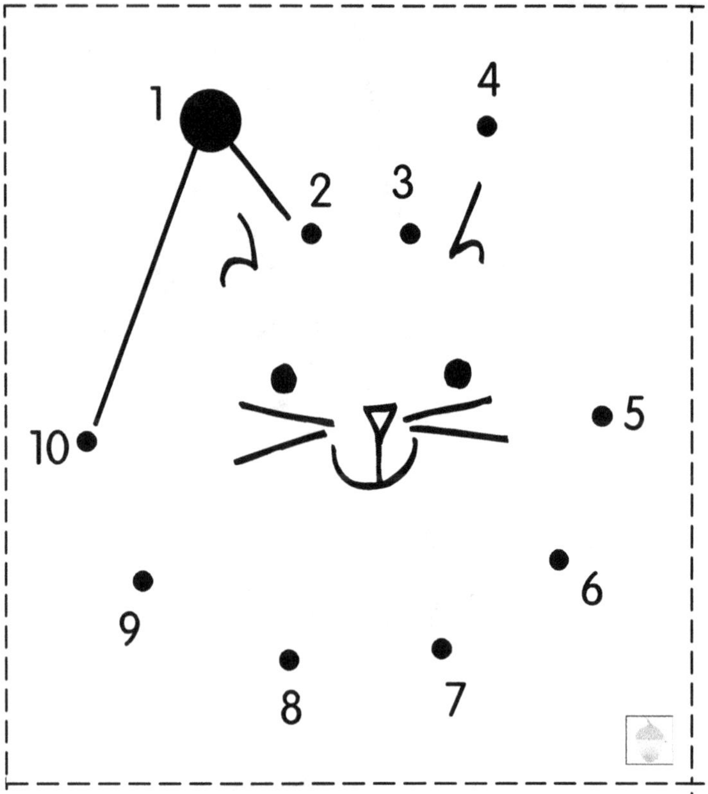

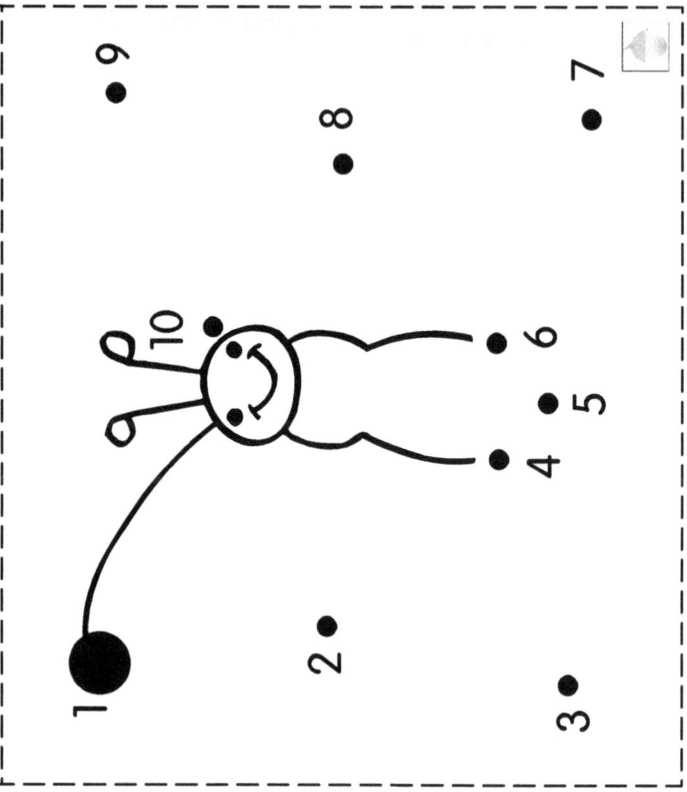

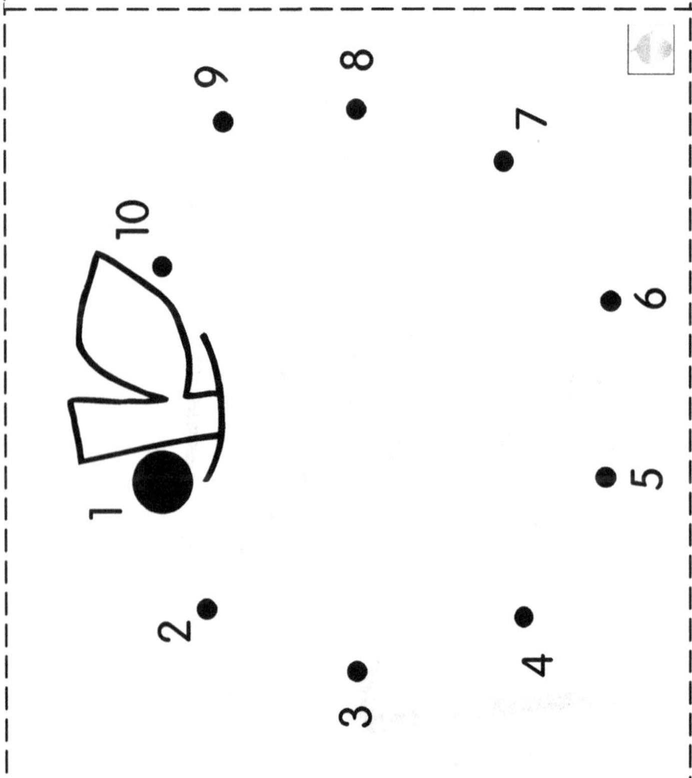

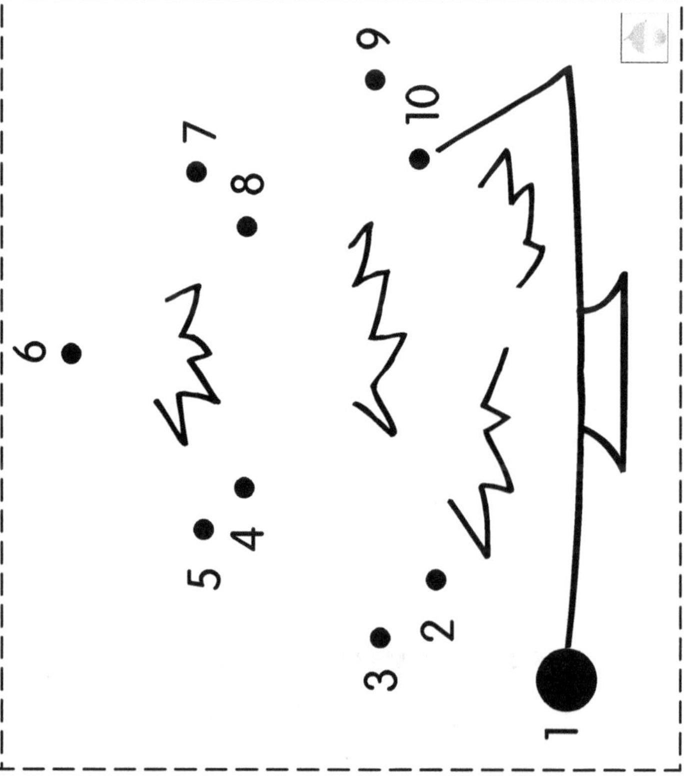

Reproduce, color, and cut apart. Enlarge if desired.

Marvelous Mouse Cover
Sequencing Numerals 1-10

Dot-to-Dot

Fabulous Fish
Sequencing Numerals 1-10

Dot-to-dot

Assembly

Reproduce, cut out, and glue a set of oak tag handles (p. 192) to a standard-size file folder. Option: Decorate each handle with a copy of the fish on this page. Reproduce, color, and cut out the "Fabulous Fish" patterns. Matching in the center, glue the fish to the inside of the folder. Glue the cover art to the front of the folder, then laminate. Laminate, then cut apart, the dot-to-dot pictures. Glue an envelope to the back of the folder to store dot-to-dot pictures, a wipe-off marker or crayon, and soft cloth. Children take out the pictures one at a time to place on the fish. Option: Reproduce, cut out, and glue the Dot-to-Dot Playing Directions to the back of the folder.

Fabulous Fish Cover
Sequencing Numerals 1-10

Dot-to-dot

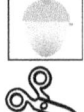

Fabulous Fish
Sequencing Numerals 1-10

Dot-to-dot

Fabulous Fish

76 LAB20145P • 35 FOLDERGAMES FOR NUMBERS • 978-1-937257-53-8 • © 2014 Little Acorn Books™

Fabulous Fish

Sequencing Numerals 1-10

Dot-to-dot

Start at the large dot on each picture.
Follow the numbers to connect the dots to make designs.

Fabulous Fish

Fabulous Fish
Sequencing Numerals 1-10

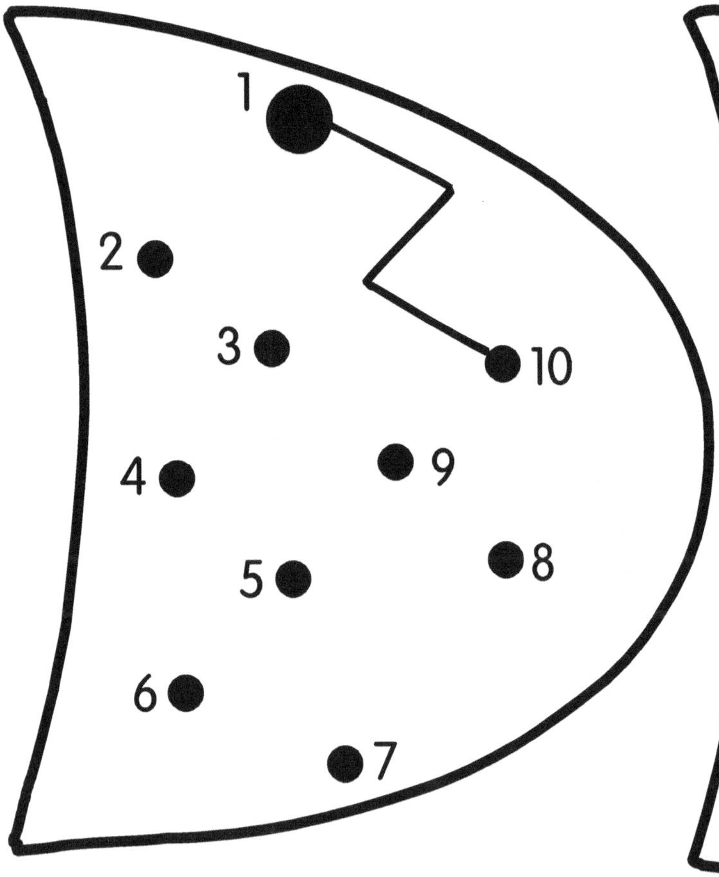

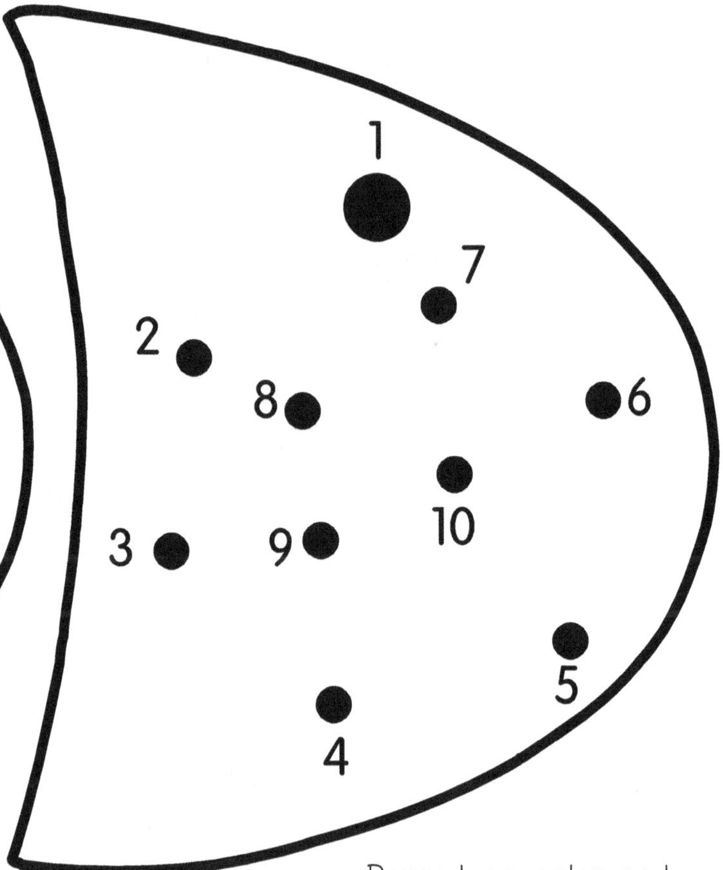

Dot-to-dot

Reproduce, color, and cut apart.

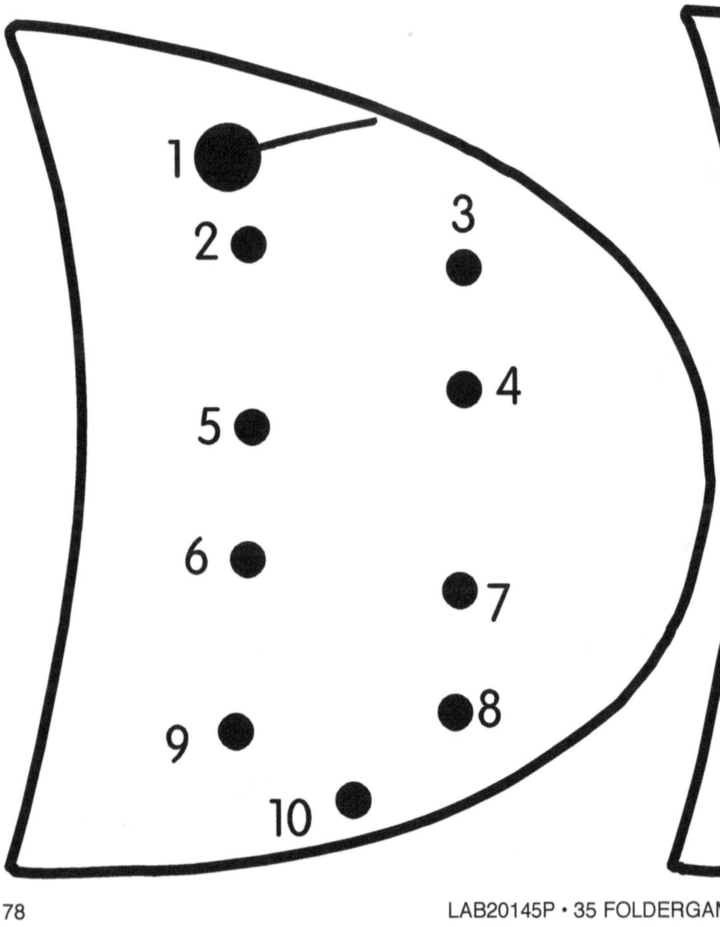

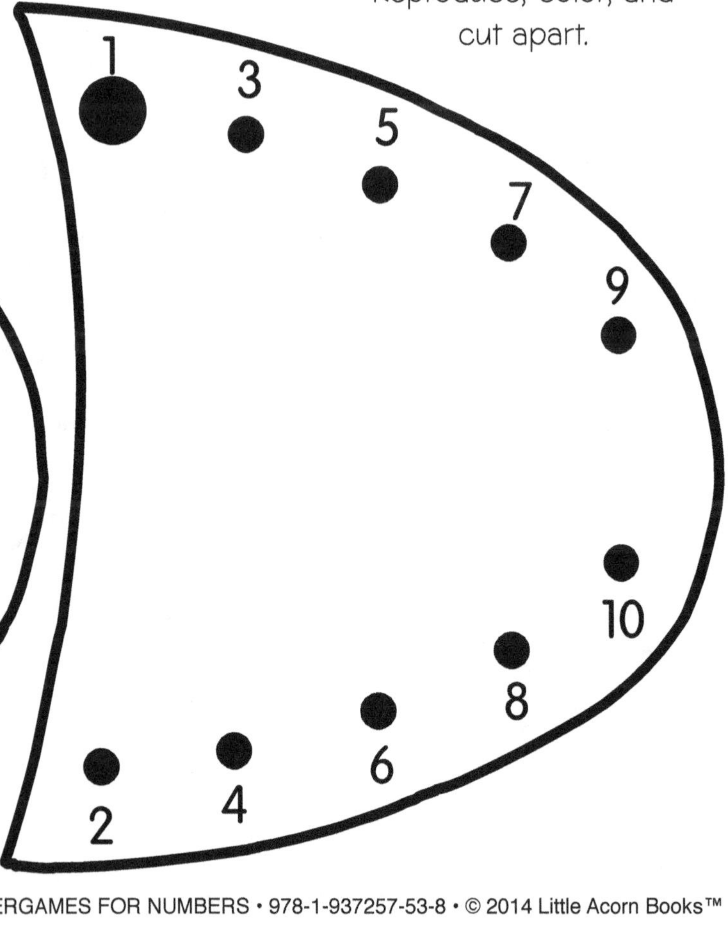

LAB20145P • 35 FOLDERGAMES FOR NUMBERS • 978-1-937257-53-8 • © 2014 Little Acorn Books™

Cheese and Crackers

Sequencing Numerals and Number Sets 1-10

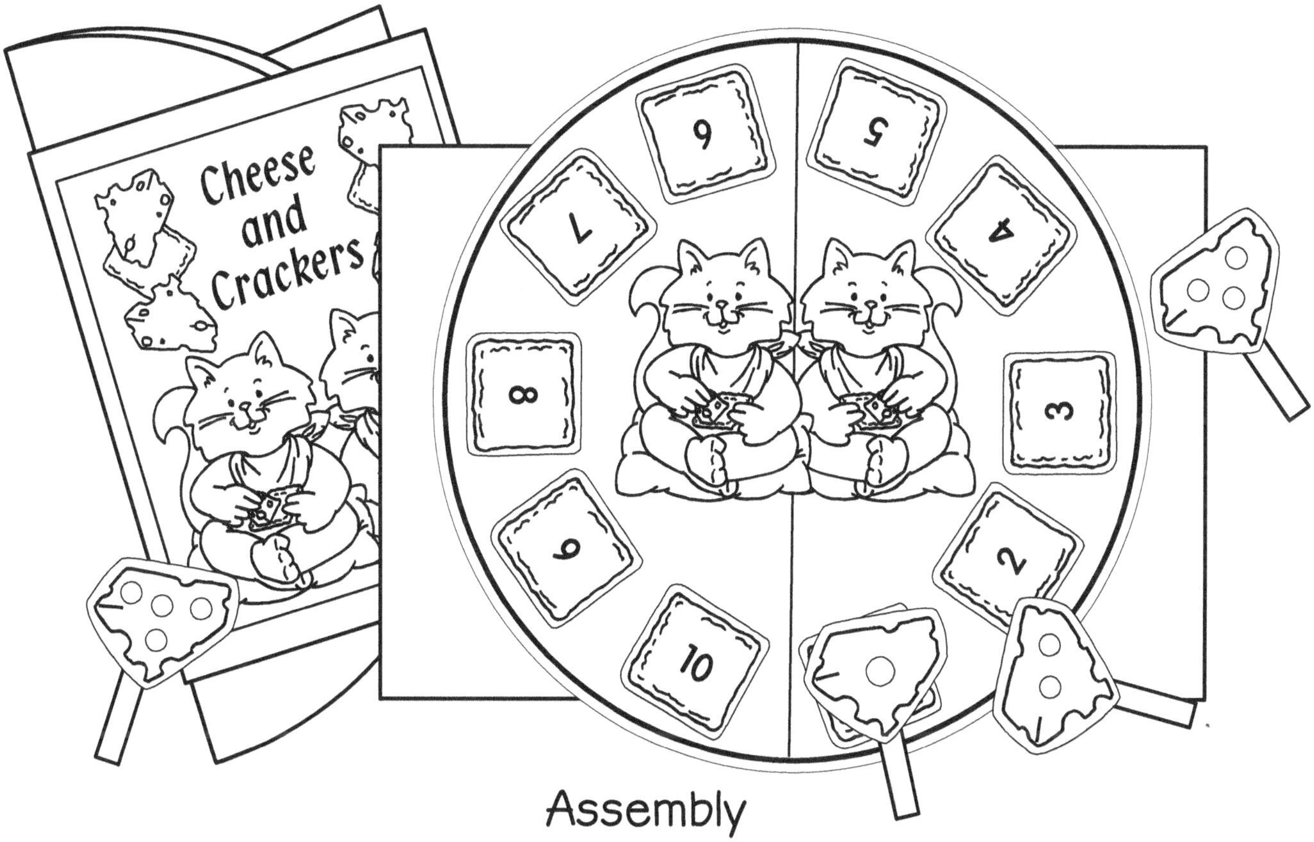

Clothespin Wheel

Assembly

Reproduce, cut out, and glue a set of oak tag handles (p. 192) to a standard-size file folder. Option: Decorate the handles with copies of the cheese wedges shown on this page. Reproduce, color, and cut out the "Cheese and Crackers" patterns. Glue the cover art to the front of the folder. Staple a resealable plastic bag to the back of the folder for clothespin storage. Glue the cheese wedges on clothespins. Meeting along the straight edges, glue the wheel quarters on a large poster board circle. Glue the wheel inside the folder, then glue the programmed crackers on the wheel in numerical order. Options: Decorate the center of the wheel. Reproduce, cut out, and glue the Clothespin Wheel directions to the back of the folder.

Directions

Reproduce, cut out, and glue the direction strip to the game board.

Pin the cheese to the matching crackers.

Cheese and Crackers Cover
Sequencing Numerals and Number Sets 1-10
Clothespin Wheel

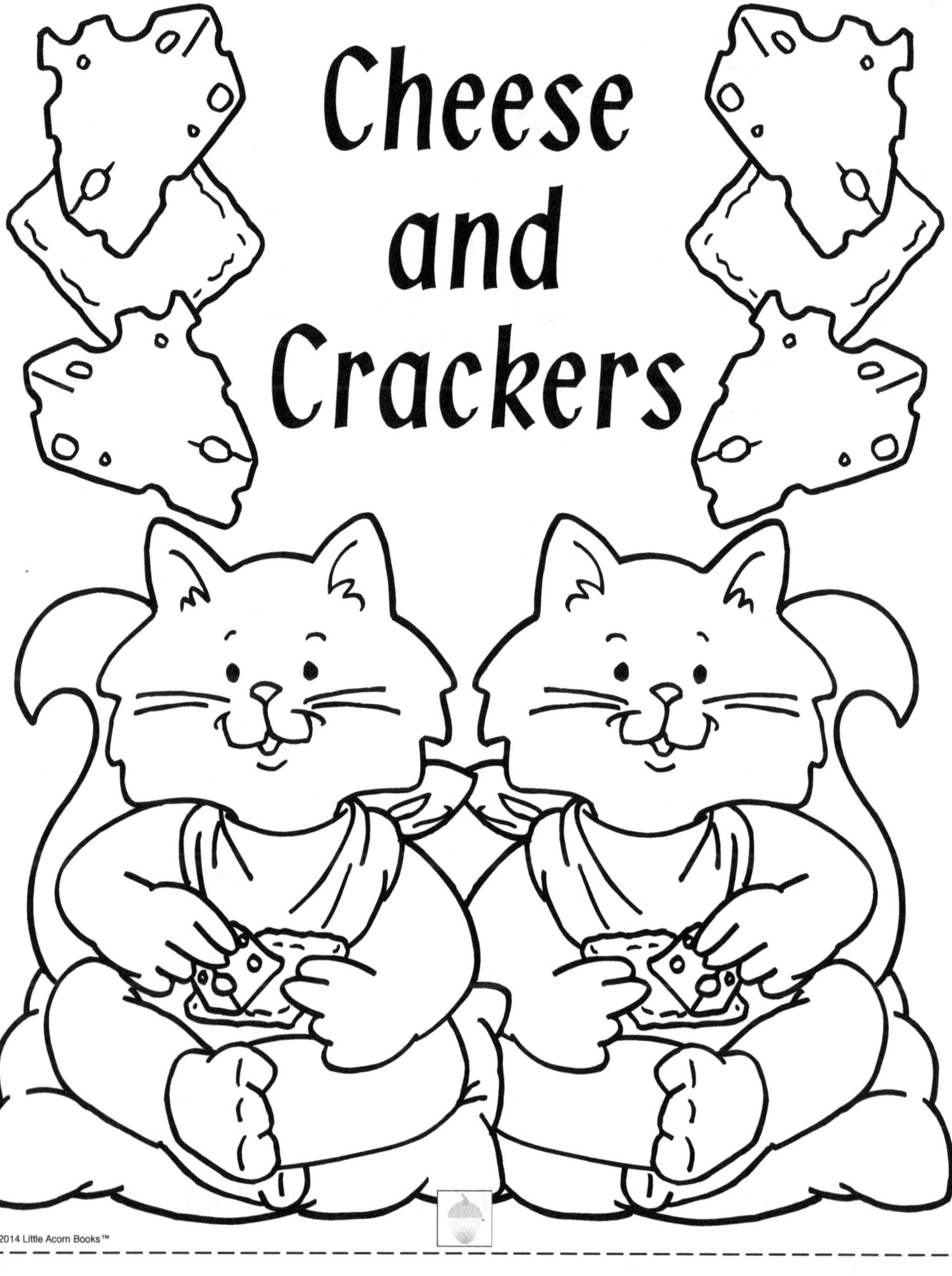

Cheese and Crackers

Cheese and Crackers
Sequencing Numerals and Number Sets 1-10

Clothespin Wheel

Reproduce, color, cut out, and glue each cheese wedge on a clothespin.

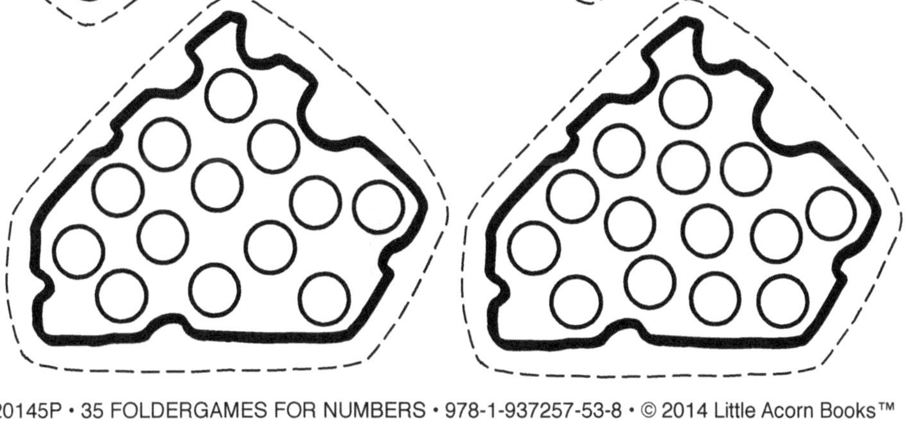

Reproduce, color, cut out, and glue the two cheese wedges on the left to clothespins for advanced skills practice.

LAB20145P • 35 FOLDERGAMES FOR NUMBERS • 978-1-937257-53-8 • © 2014 Little Acorn Books™

Cheese and Crackers
Sequencing Numerals and Number Sets 1-10

Clothespin Wheel

Reproduce, color, cut out, and glue each cracker around an assembled clothespin wheel.

Panda Lotto
Matching Number Sets and Number Words 1-10

Lotto

Assembly

Reproduce, cut out, and glue a set of oak tag handles (p. 192) to a standard-size file folder. Option: Decorate each handle with a copy of the pandas shown above. Reproduce, color, and cut out the "Panda Lotto" patterns. Glue the cover art to the front of the folder. Mount Lotto boards on colored construction paper, then laminate. Glue two envelopes to the inside of the folder for Lotto board, token, and card storage. Option: Reproduce, cut out, and glue the Lotto Playing Directions to the back of the folder. Note: Wipe-off markers or crayons can also be used to color in matches.

Tokens

Reproduce, color, and cut out tokens for players to place on Lotto boards.

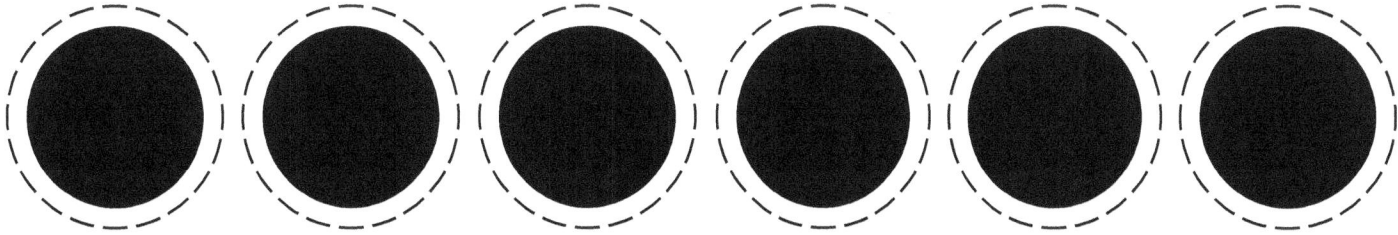

Panda Lotto Cover
Matching Number Sets and Number Words 1-10

Panda Lotto

Matching Number Sets and Number Words 1-10

Lotto

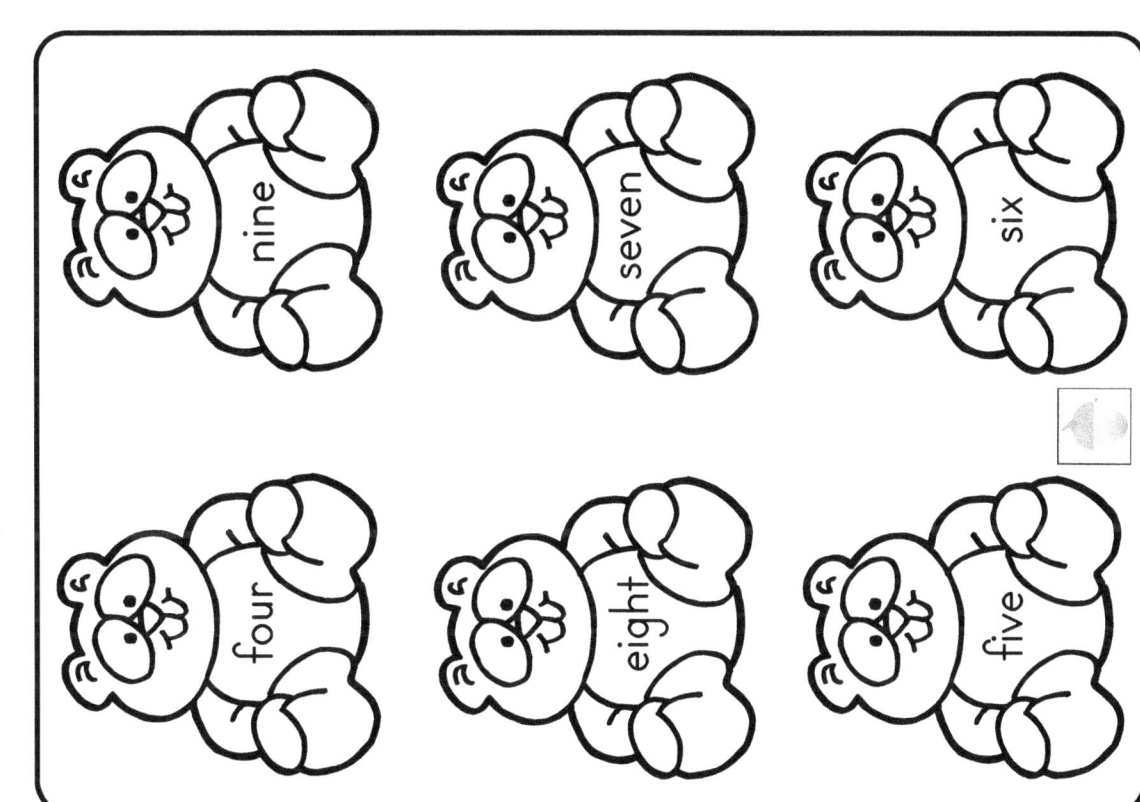

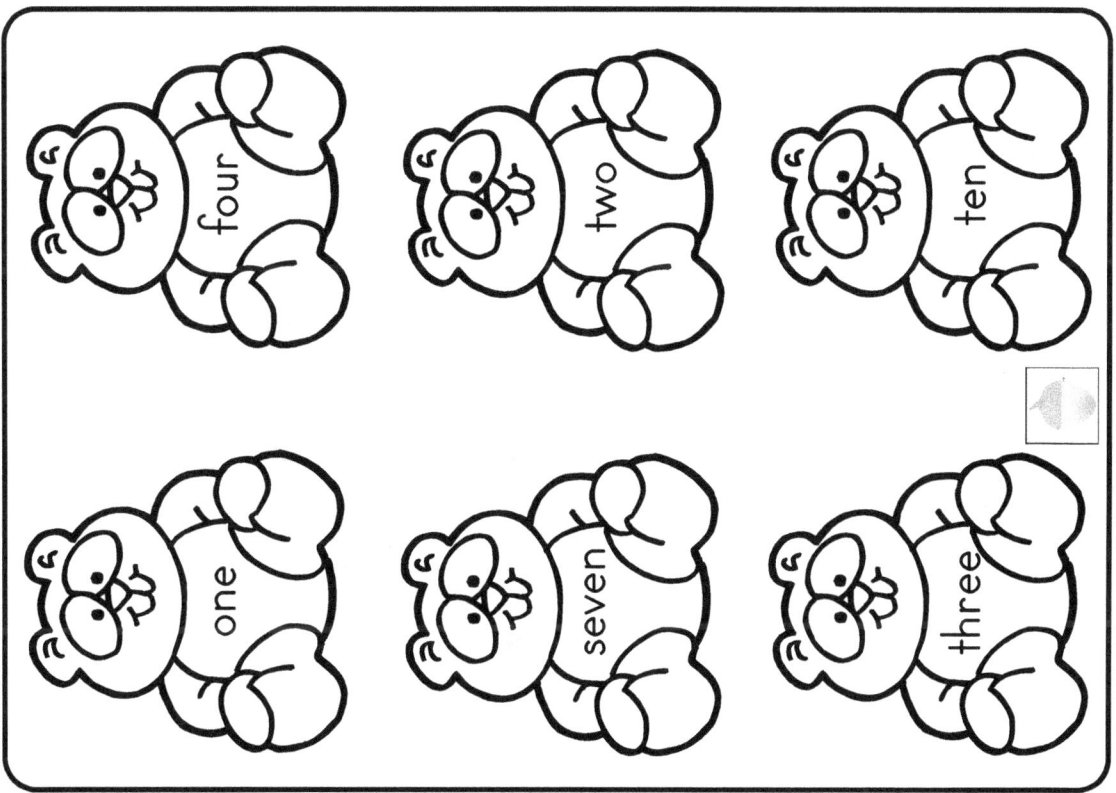

Panda Lotto
Matching Number Sets and Number Words 1-10

Lotto

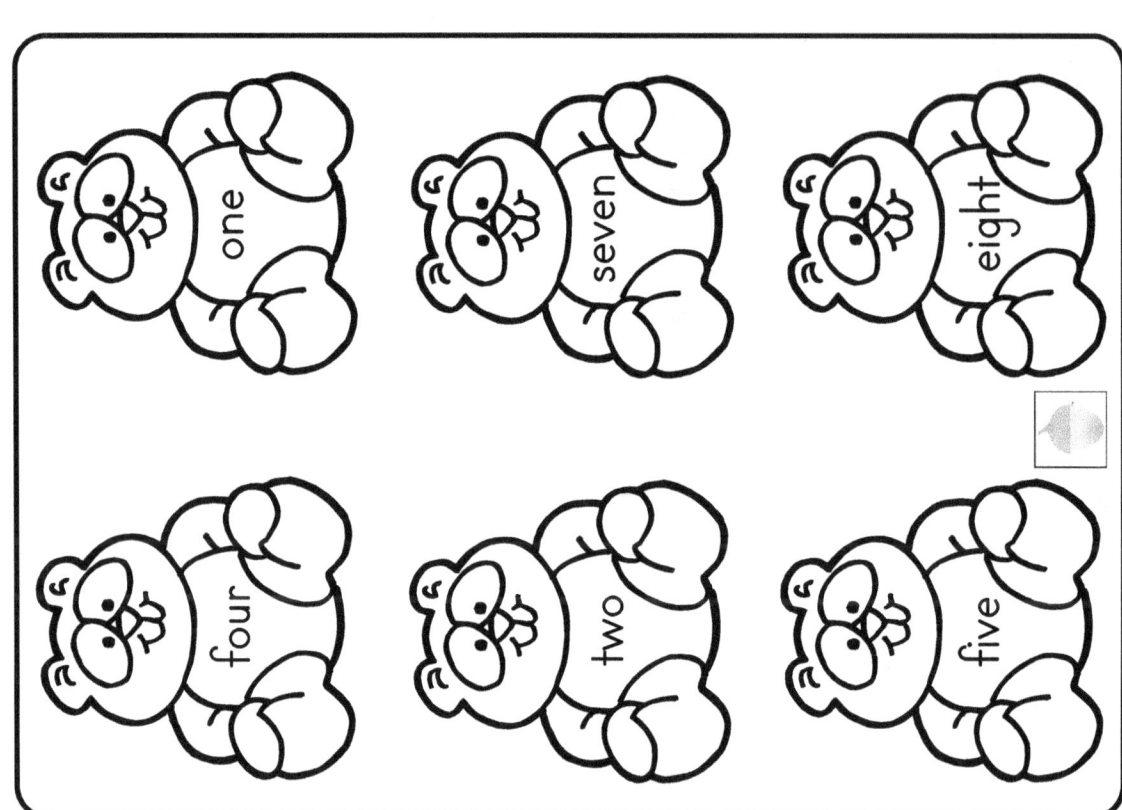

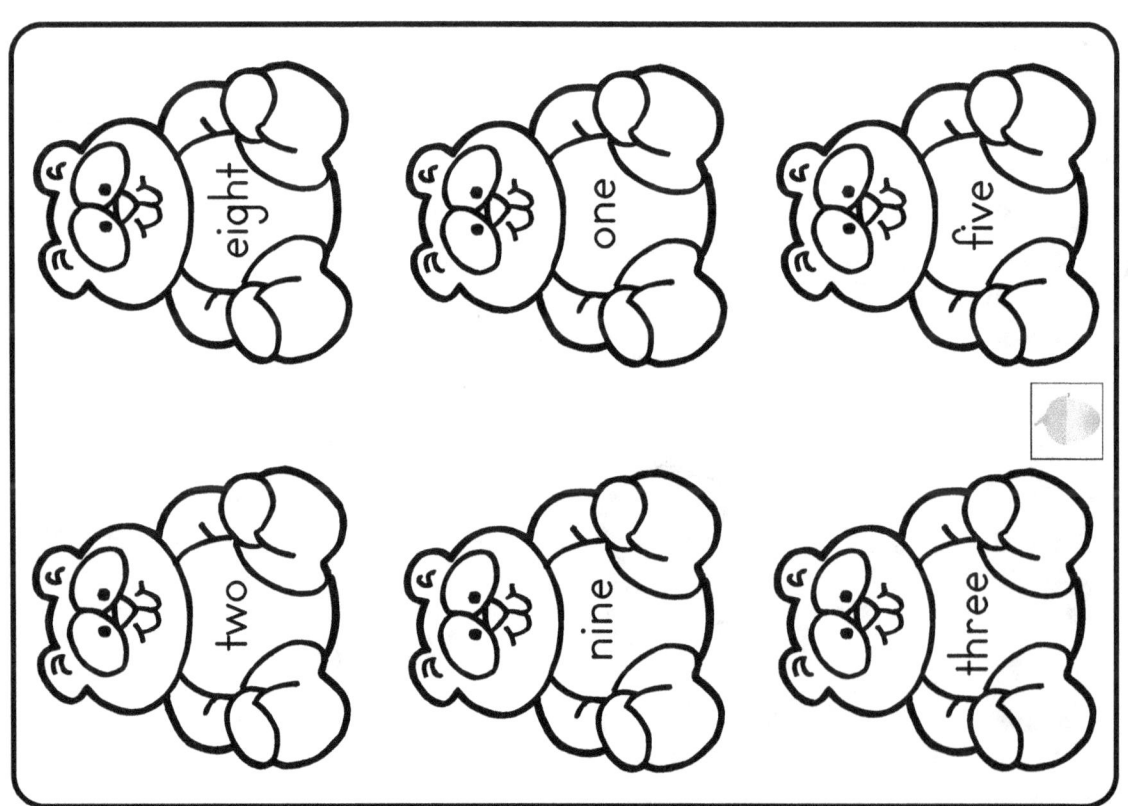

Panda Lotto

Matching Number Sets and Number Words 1-10

Lotto

Reproduce, color, and cut apart two sets of cards.

Tumbling Turtles
Matching Number Sets and Number Words 1-10

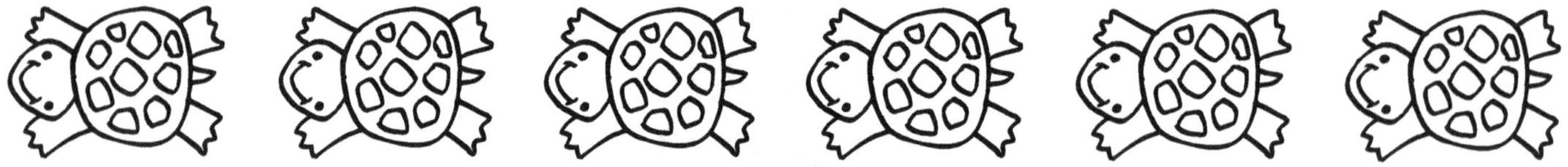

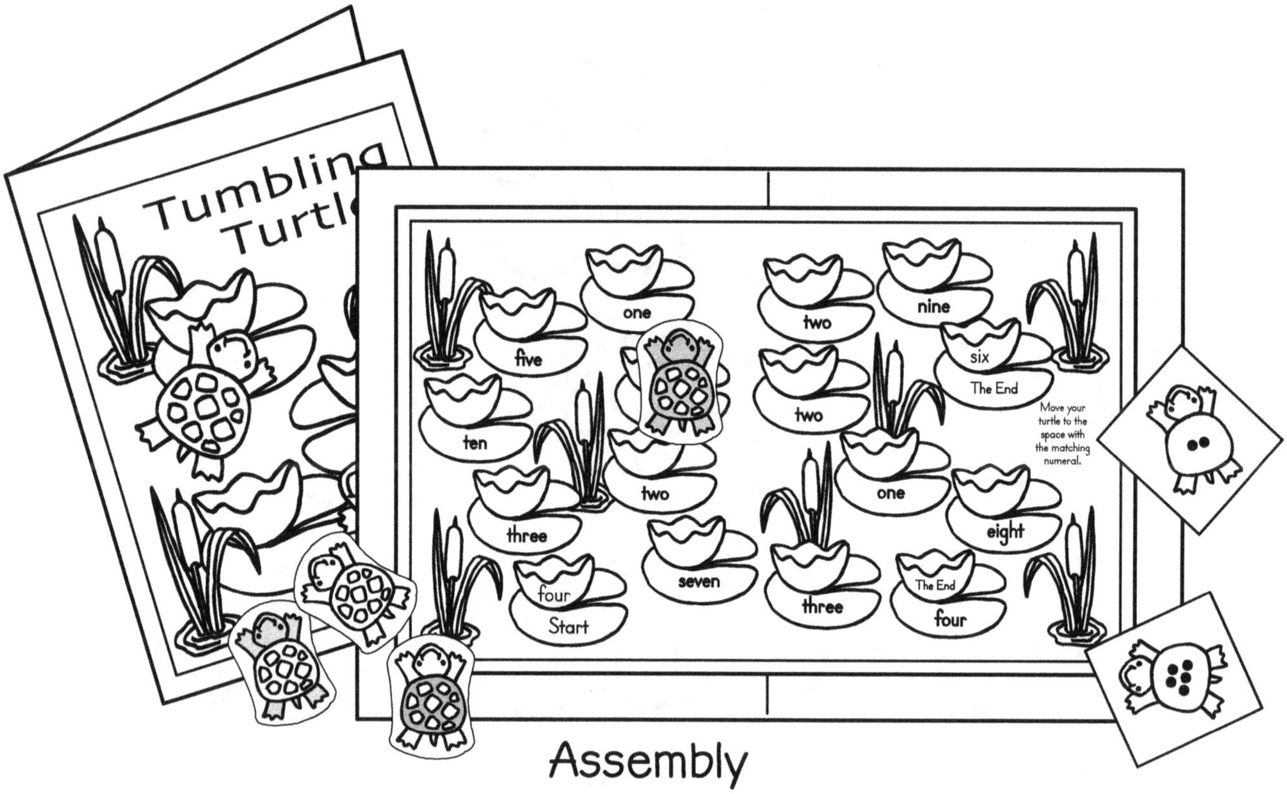

Trail Game

Assembly

Reproduce, cut out, and glue a set of oak tag handles (p. 192) to a standard-size file folder. Option: Decorate each handle with a copy of the turtles on this page. Reproduce, color, and cut out the "Tumbling Turtles" patterns. Matching in the center, glue the game board patterns to the inside of the folder. Glue the cover art to the front of the folder. Decorate the border around the game board, then laminate. Reproduce, color, laminate, then cut out the pawns and two sets of game cards. Glue an envelope to the back of the folder for pawn and game card storage. Option: Reproduce, cut out, and glue the Trail Game Playing Directions to the back of the folder.

Pawns
Reproduce, color each turtle a different color, and cut out.

Tumbling Turtles Cover
Matching Number Sets and Number Words 1-10

Trail Game

Tumbling Turtles

Tumbling Turtles
Matching Number Sets and Number Words 1-10

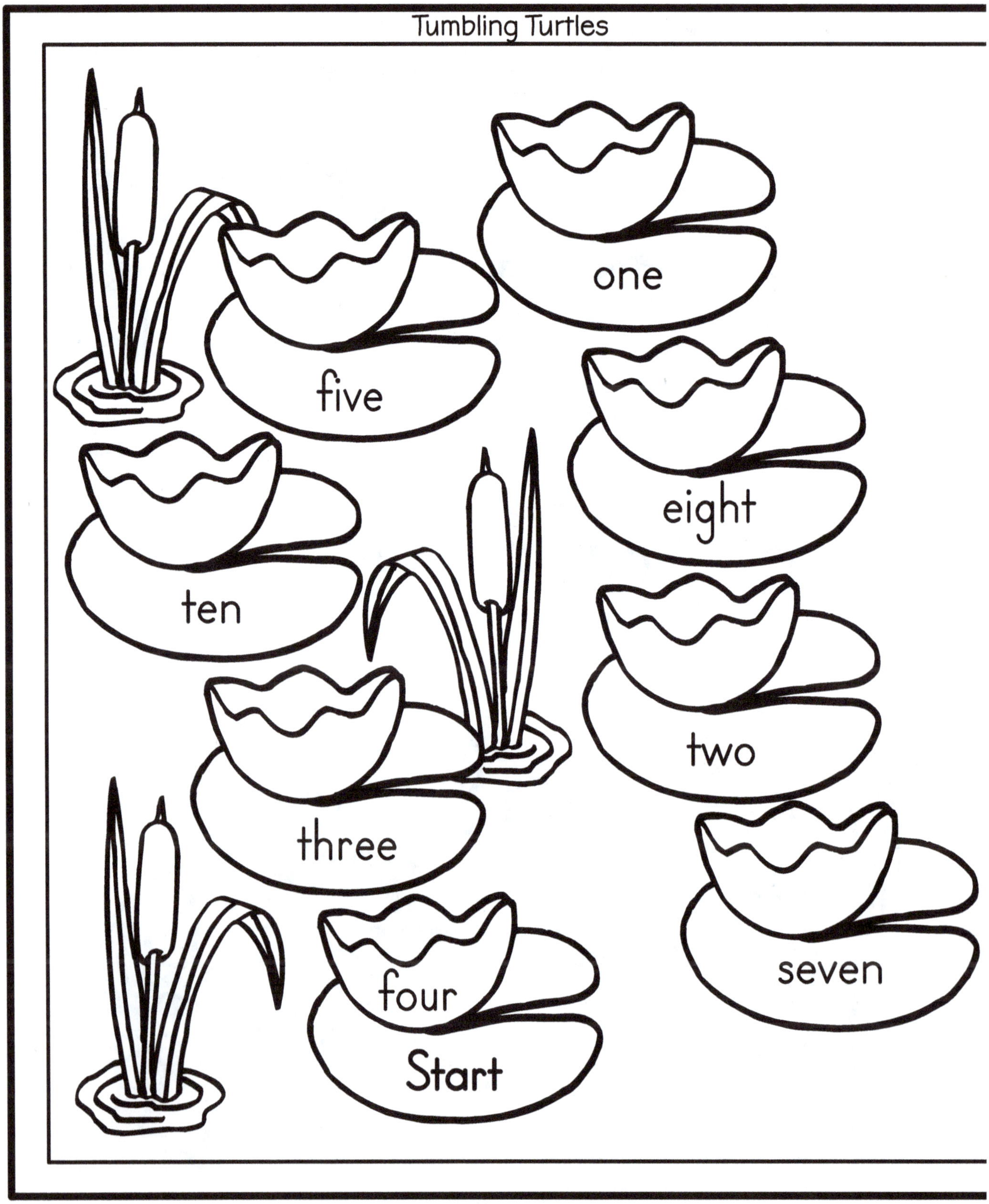

Trail Game

Tumbling Turtles

- one
- five
- eight
- ten
- two
- three
- seven
- four Start

Trail Game

Tumbling Turtles

Matching Number Sets and Number Words 1-10

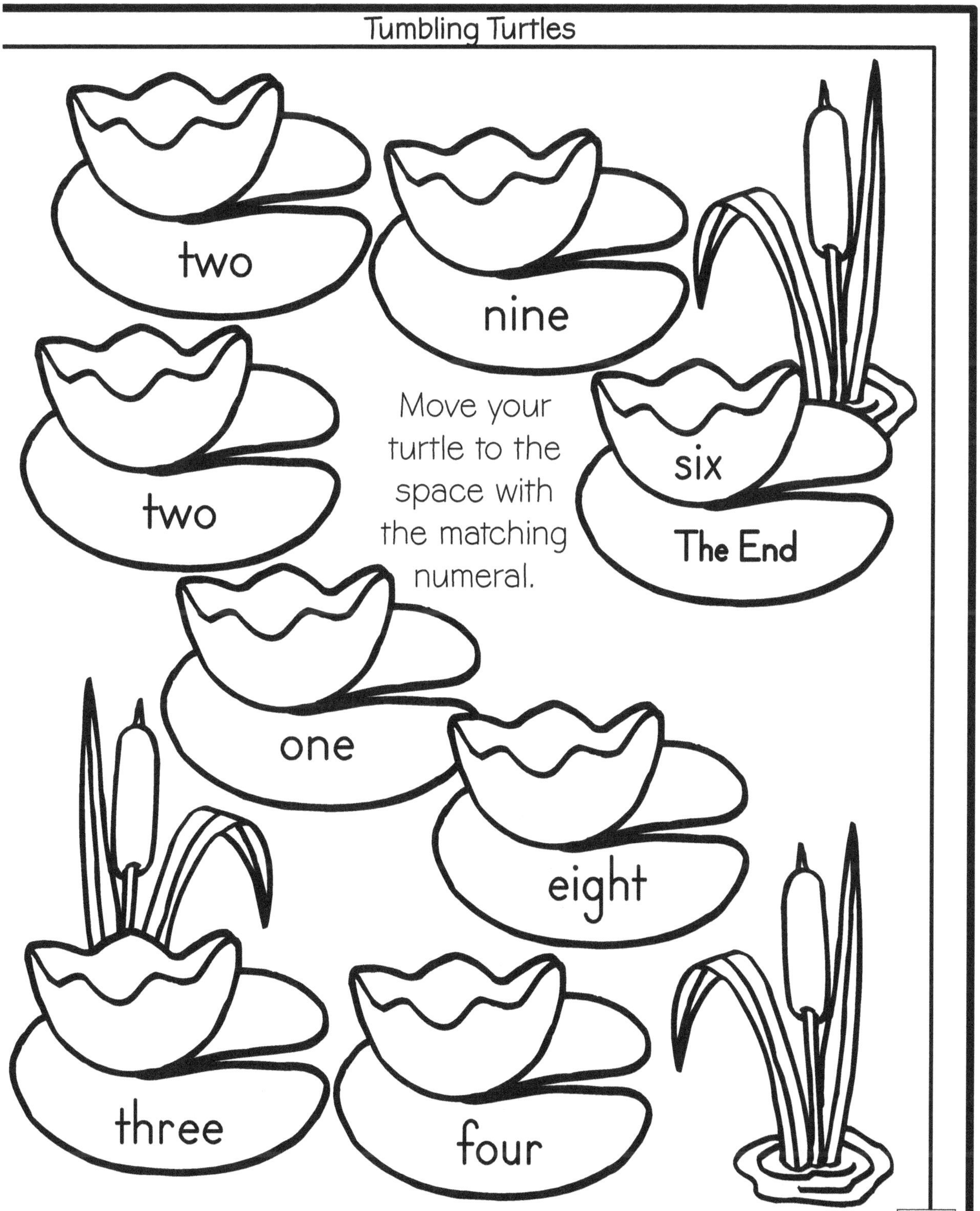

Tumbling Turtles
two
nine
two
Move your turtle to the space with the matching numeral.
six
The End
one
eight
three
four
© 2014 Little Acorn Books™

Tumbling Turtles
Matching Number Sets and Number Words 1-10

Trail Game

Reproduce, color, and cut apart two sets of cards.

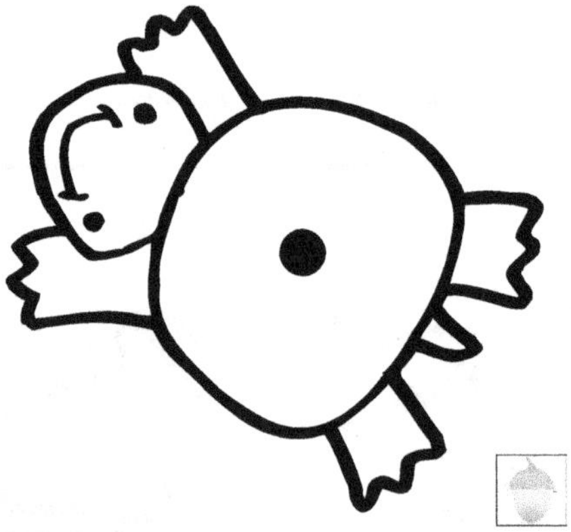

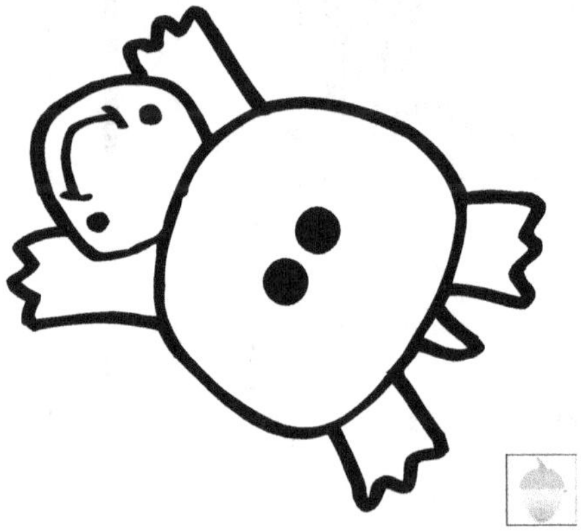

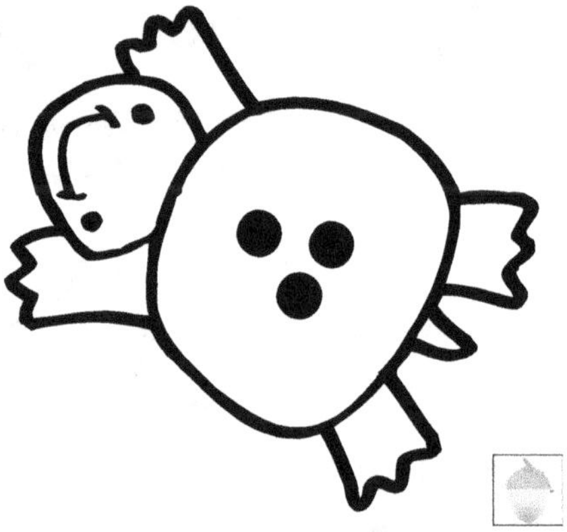

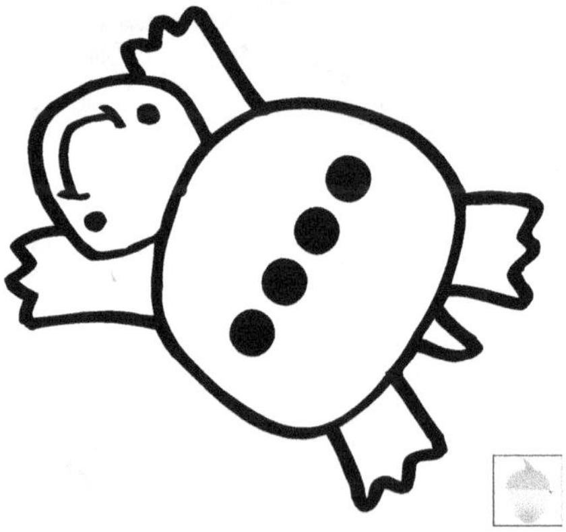

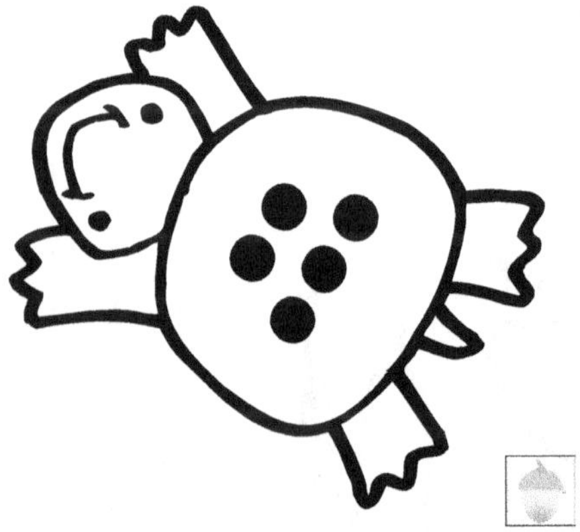

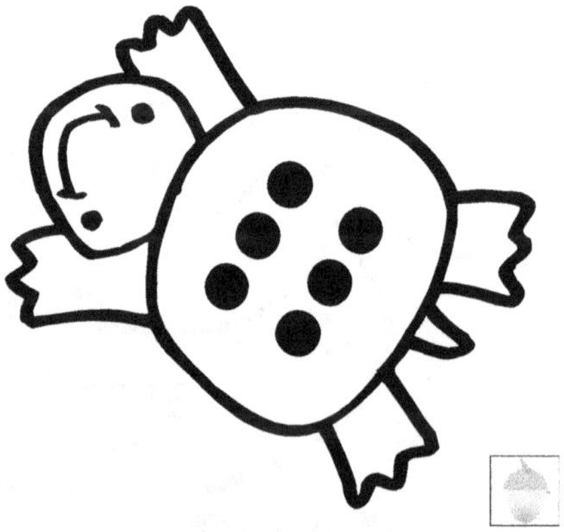

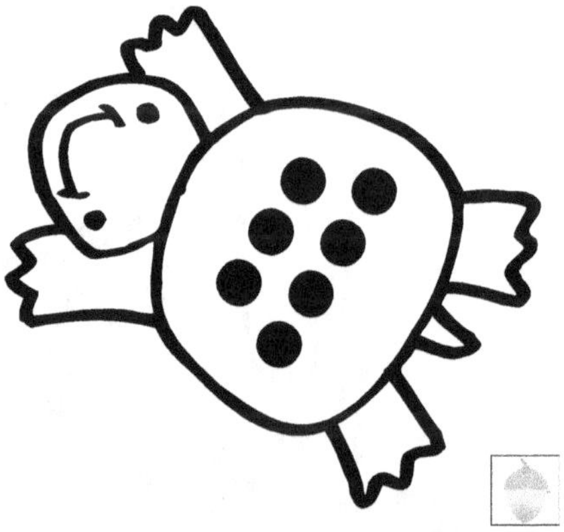

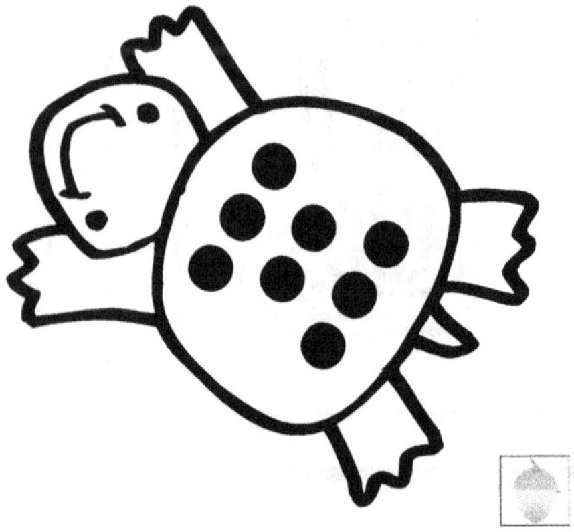

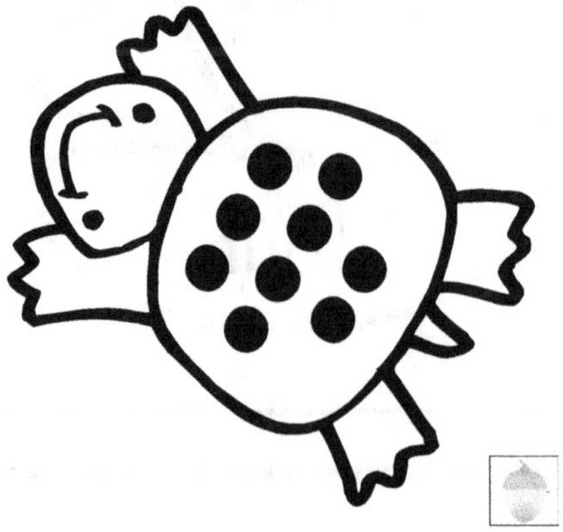

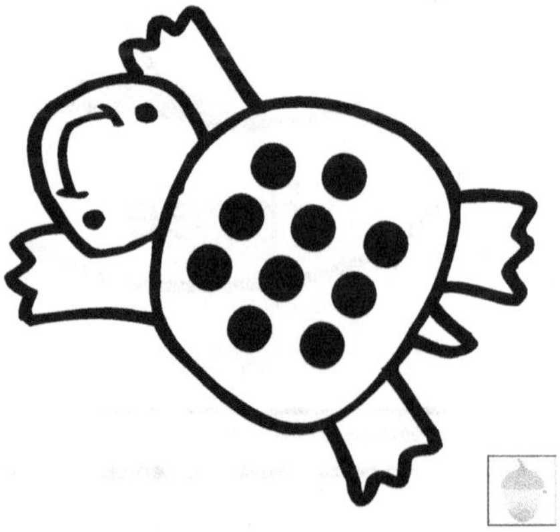

Koala King Bingo
Matching Number Sets and Number Words 1-10

Bingo

Assembly

Reproduce, cut out, and glue a set of oak tag handles (p. 192) to a standard-size file folder. Option: Decorate each handle with a copy of the crowns shown above. Reproduce, color, and cut out the "Koala King Bingo" patterns. Mount Bingo boards on colored construction paper, then laminate. Glue the cover art to the front of the folder. Glue two envelopes to the inside of the folder for Bingo board, token, and card storage. Option: Reproduce, cut out, and glue the Bingo Playing Directions to the back of the folder. Note: Wipe-off markers or crayons can be also be used to color in matches.

Tokens
Reproduce crown tokens for players to use as markers.

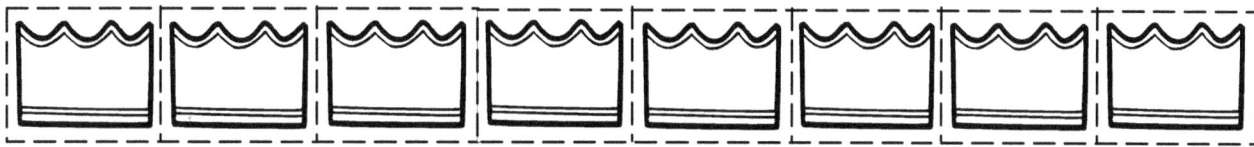

Koala King Bingo Cover
Matching Number Sets and Number Words 1-10

Bingo

Koala King BINGO

Koala King Bingo

Matching Number Sets and Number Words 1-10

Koala King Bingo

K	O	A	L	A
three	eight	six	nine	ten
ten	five	nine	one	six
seven	four	FREE SPACE	six	three
seven	two	eight	two	one
four	ten	five	seven	ten

Koala King Bingo

K	O	A	L	A
four	five	one	two	ten
six	eight	seven	five	four
six	three	FREE SPACE	three	seven
four	nine	four	six	two
nine	ten	three	one	eight

Koala King Bingo
Matching Number Sets and Number Words 1-10

Bingo

Koala King Bingo

K	O	A	L	A
eight	three	ten	nine	six
nine	six	five	seven	four
seven	four	FREE SPACE	five	one
one	ten	eight	nine	nine
three	seven	two	two	eight

Koala King Bingo

K	O	A	L	A
eight	five	four	five	seven
two	one	six	seven	nine
ten	seven	FREE SPACE	three	four
five	three	ten	nine	two
one	six	nine	eight	five

Koala King Bingo
Matching Number Sets and Number Words 1-10

Bingo

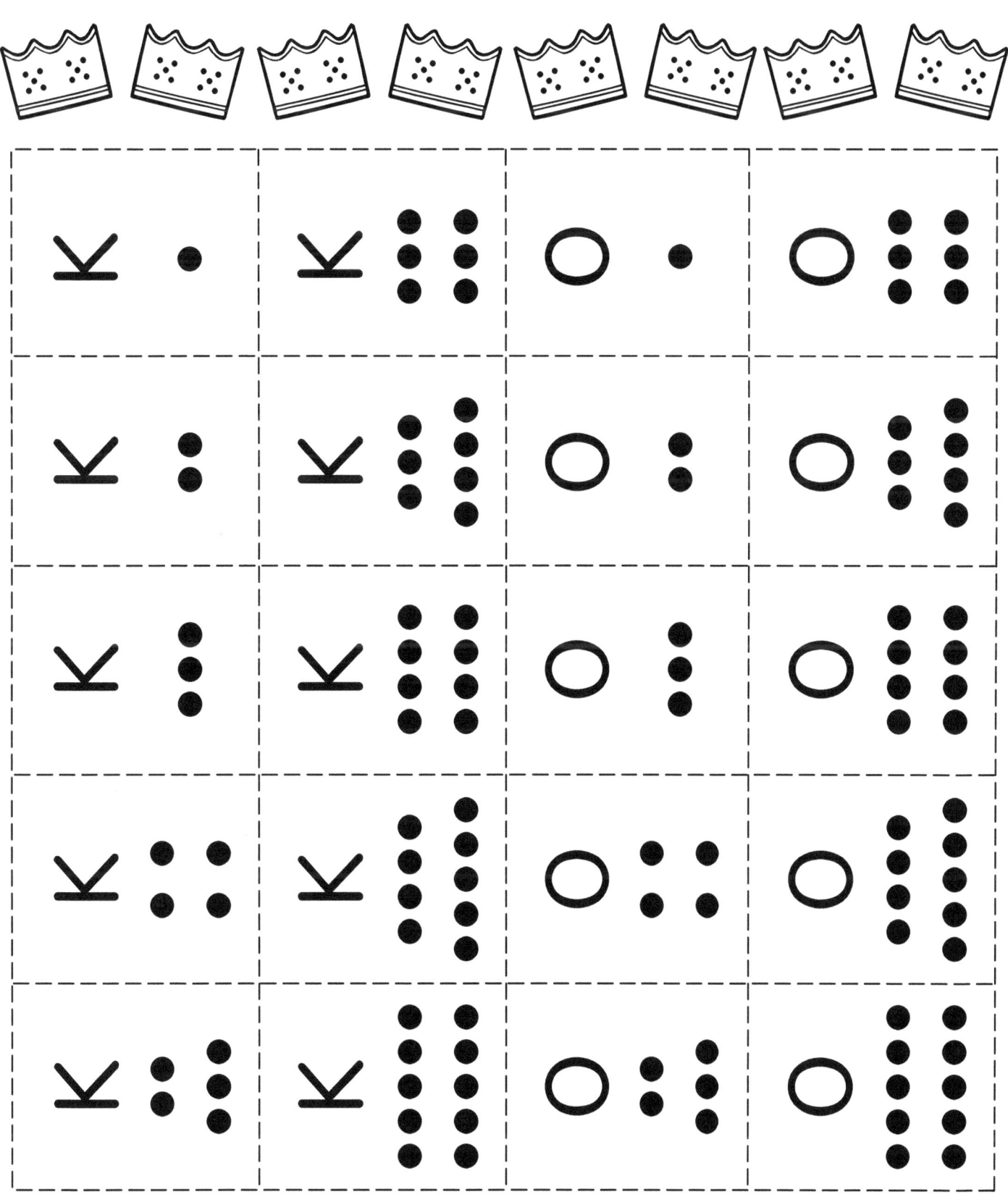

Reproduce, color, laminate, and cut apart

Koala King Bingo
Matching Number Sets and Number Words 1-10

Bingo

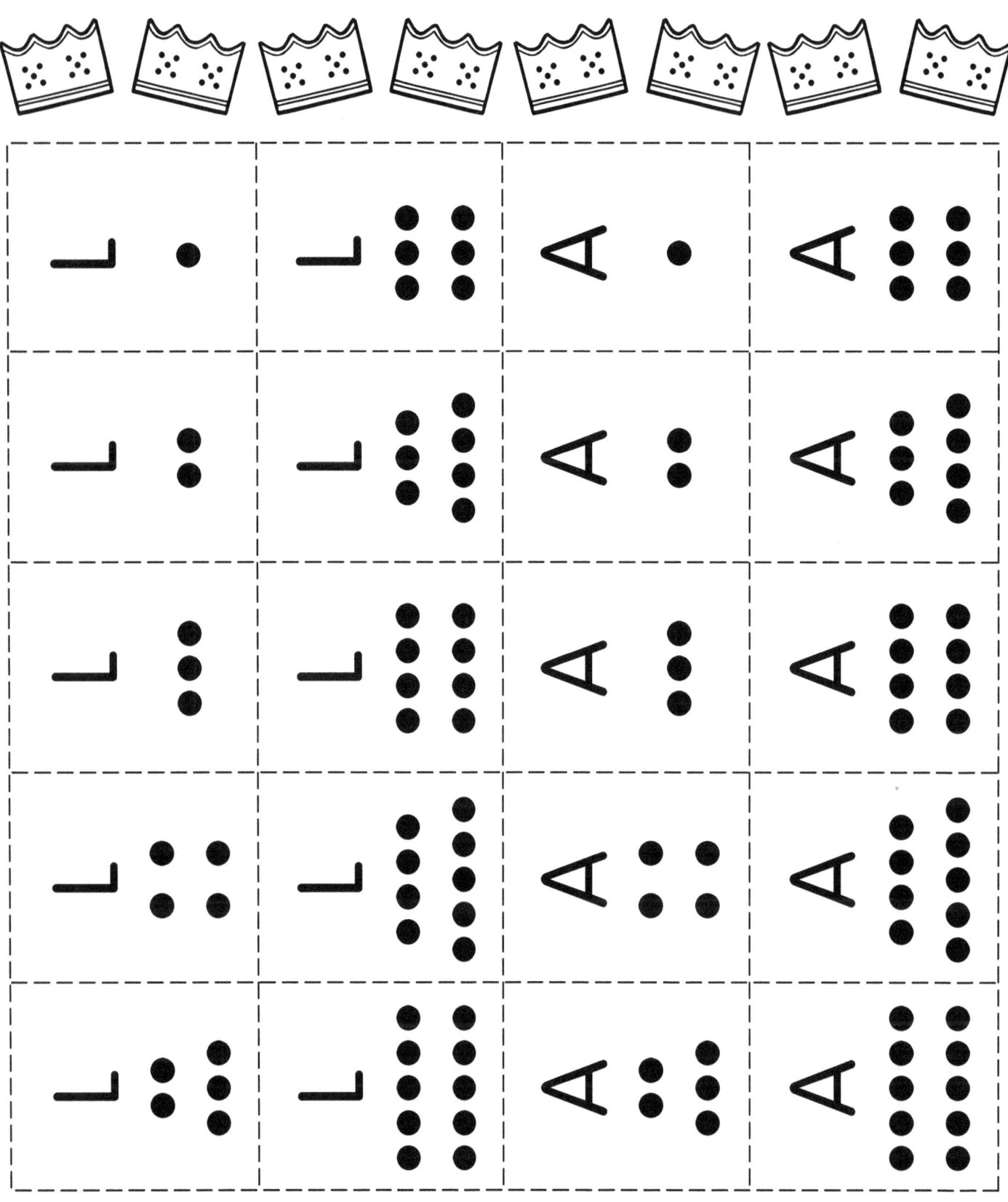

Reproduce, color, laminate, and cut apart.

Rubber Duck Dominoes
Matching Number Sets and Number Words 1-10

Dominoes

Assembly

Reproduce, cut out, and glue a set of oak tag handles (p. 192) to a standard-size file folder. Option: Decorate the handles with copies of the ducks shown on this page. Reproduce, color, and cut out the "Rubber Duck Dominoes" patterns. Glue the cover art to the front of the folder. Decorate the inside of the folder, then laminate. Reproduce, color, laminate, then cut apart the dominoes. Glue an envelope to the back of the folder to store dominoes. Option: Reproduce, cut out, and glue the Dominoes Playing Directions to the back of the folder.

Rubber Duck Dominoes Cover
Matching Number Sets and Number Words 1-10

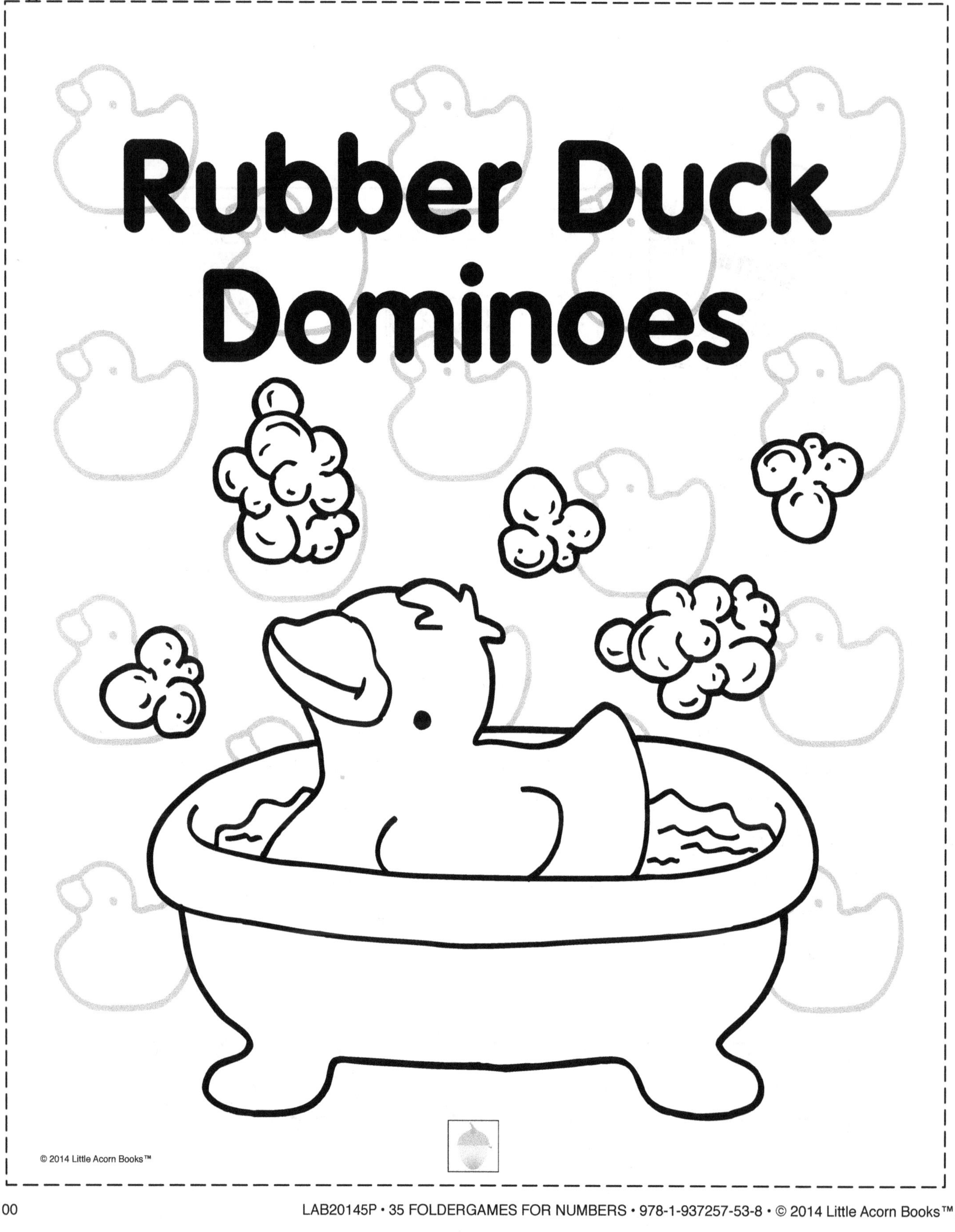

Dominoes

Rubber Duck Dominoes

© 2014 Little Acorn Books™

Rubber Duck Dominoes

Matching Number Sets and Number Words 1-10

Dominoes

Rubber Duck Dominoes
Matching Number Sets and Number Words 1-10

Dominoes

Rubber Duck Dominoes
Matching Number Sets and Number Words 1-10

Dominoes

Cupcake Lotto Cover
Matching Numerals and Number Words 1-10

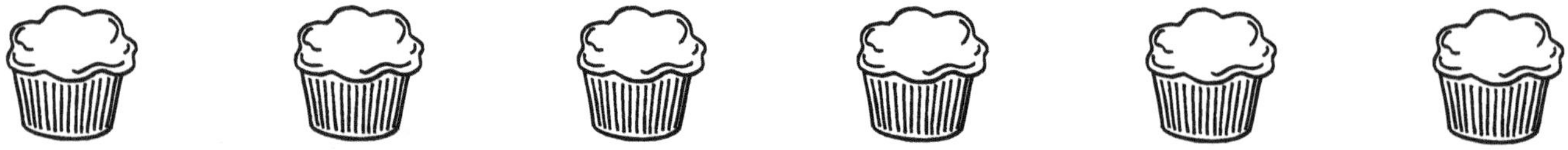

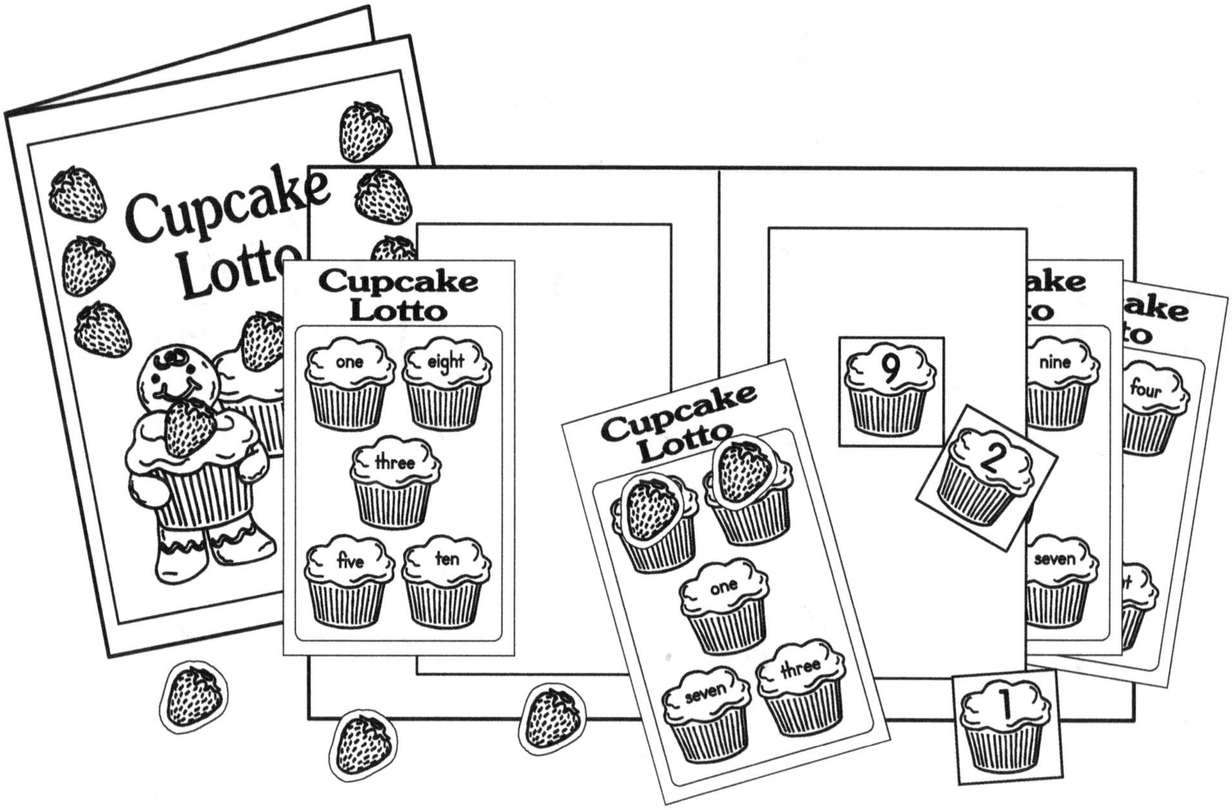

Lotto

Assembly

Reproduce, cut out, and glue a set of oak tag handles (p. 192) to a standard-size file folder. Option: Decorate each handle with a copy of the cupcakes shown above. Reproduce, color, and cut out the "Cupcake Lotto" patterns. Glue the cover art to the front of the folder. Mount Lotto boards on colored construction paper, then laminate. Glue two envelopes to the inside of the folder for Lotto board, token, and card storage. Option: Reproduce, cut out, and glue the Lotto Playing Directions to the back of the folder. Note: Wipe-off markers or crayons can also be used to color in matches.

Tokens
Reproduce, color, and cut out tokens for players to place on Lotto boards.

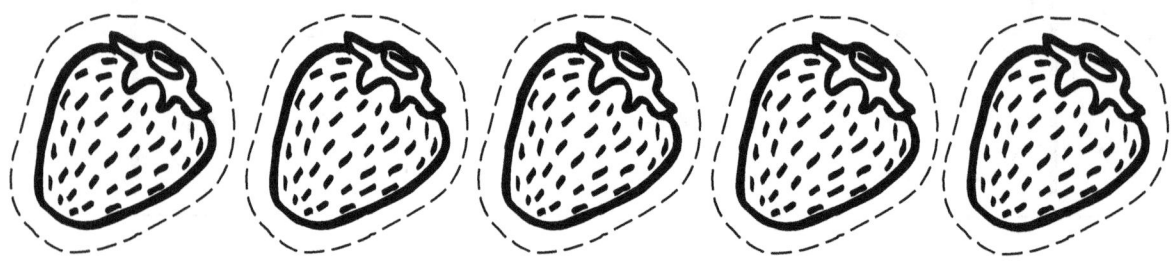

Cupcake Lotto Cover
Matching Numerals and Number Words 1-10

Lotto

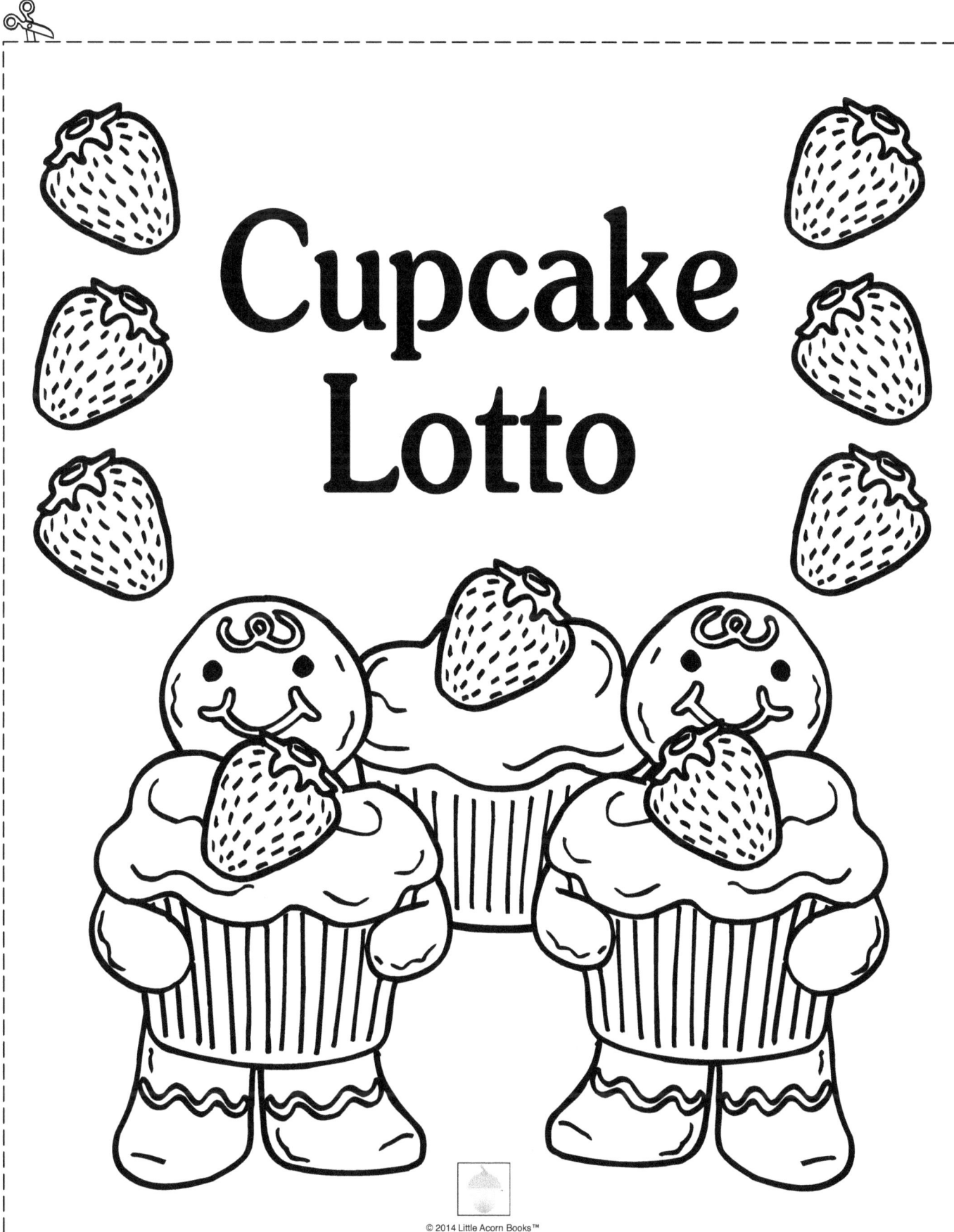

Cupcake Lotto
Matching Numerals and Number Words 1-10

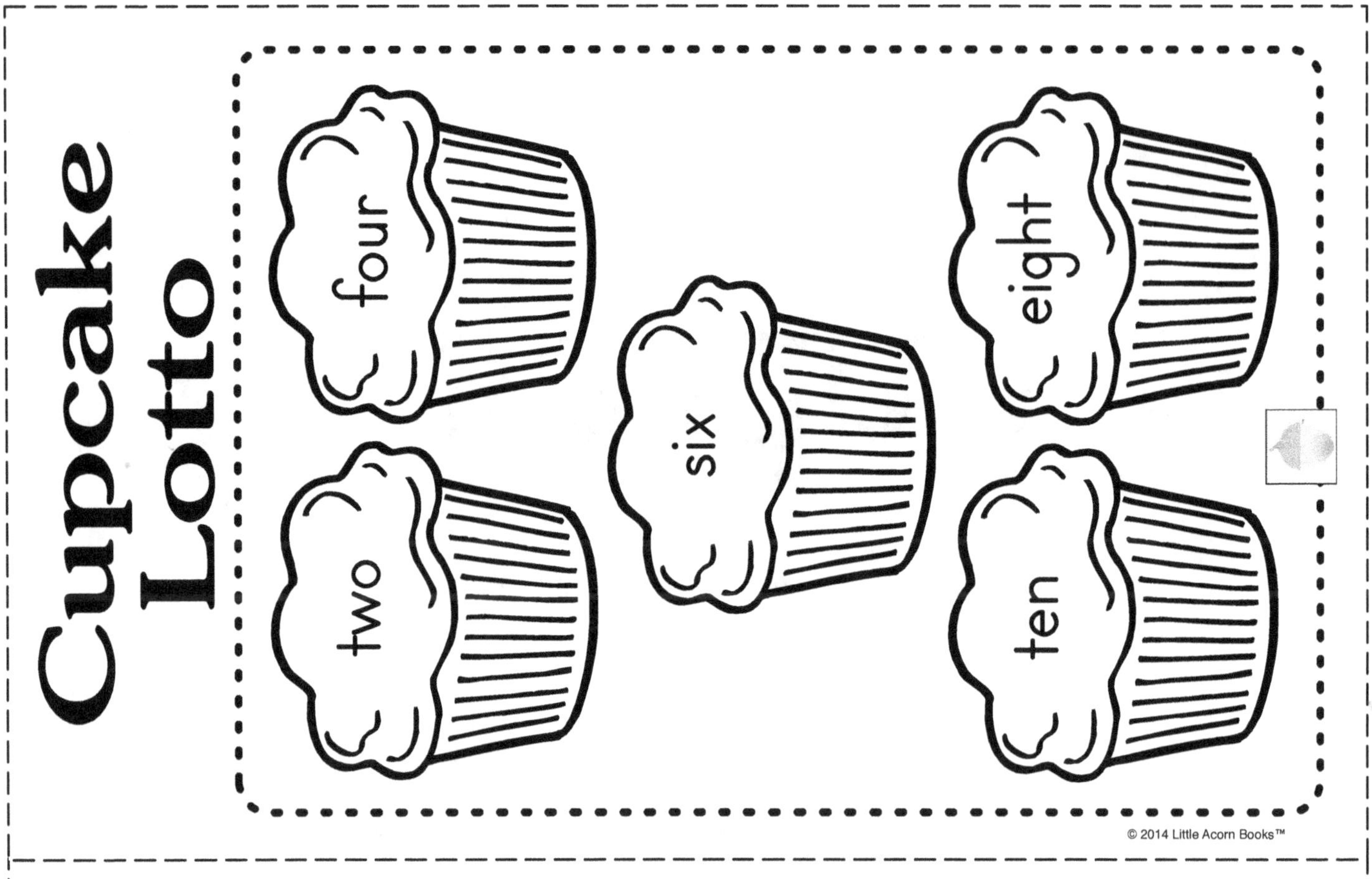

Lotto

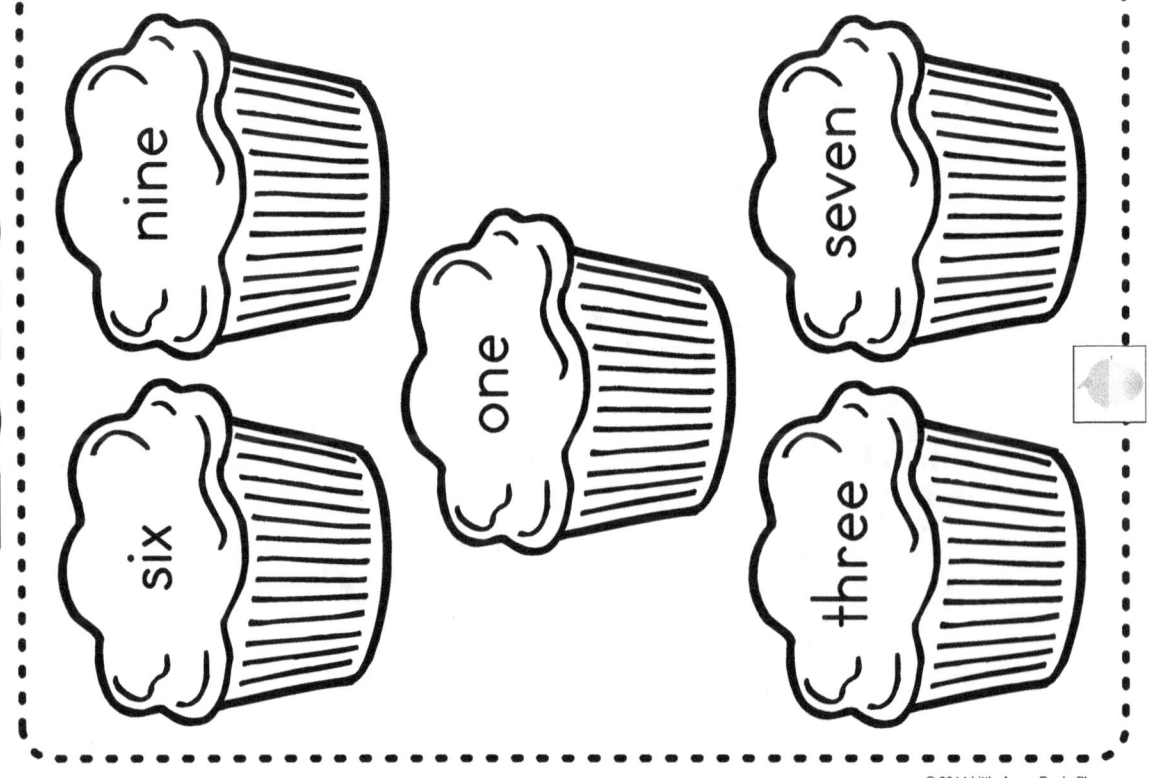

Cupcake Lotto
Matching Numerals and Number Words 1-10

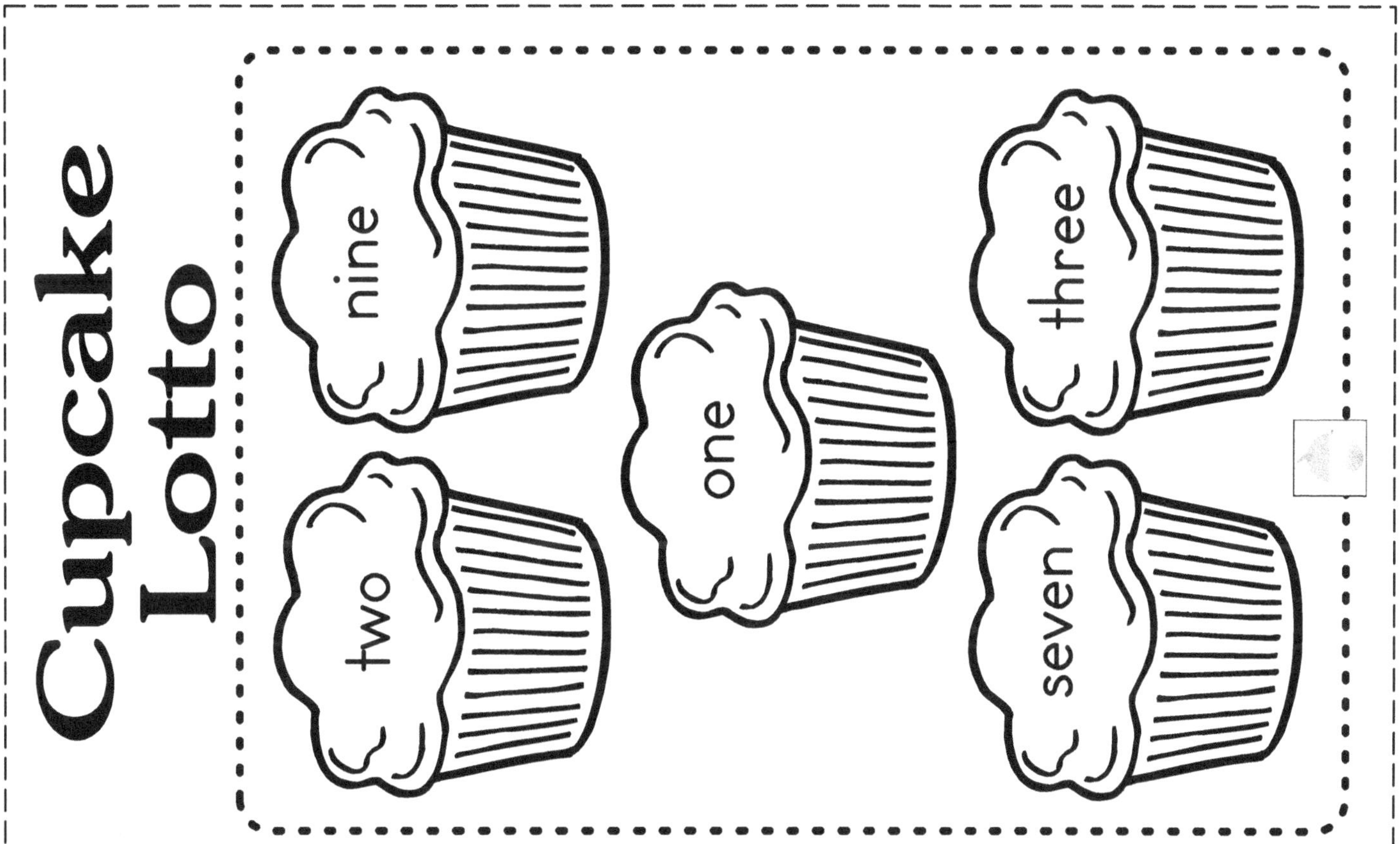

Lotto

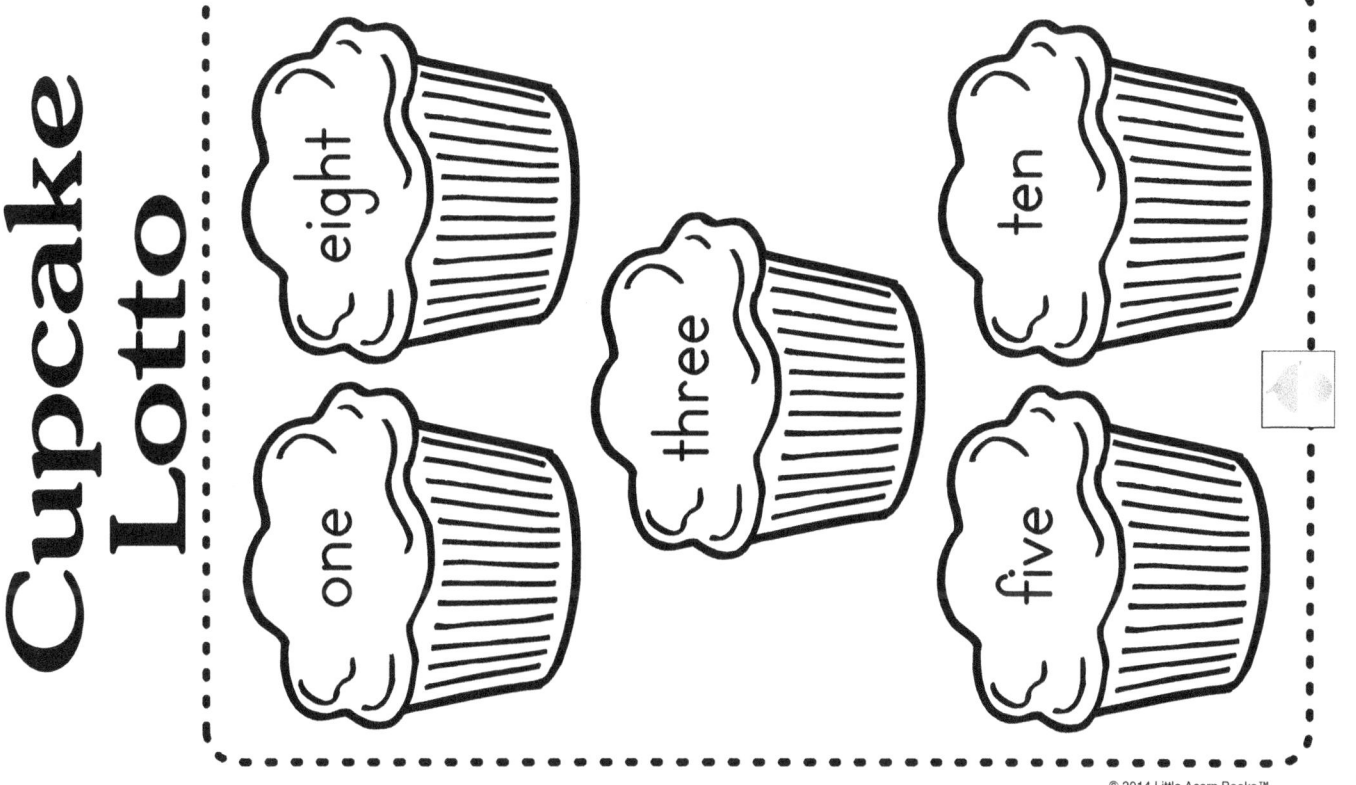

Cupcake Lotto

Lotto

Matching Numerals and Number Words 1-10

Reproduce, color, and cut apart two sets of cards.

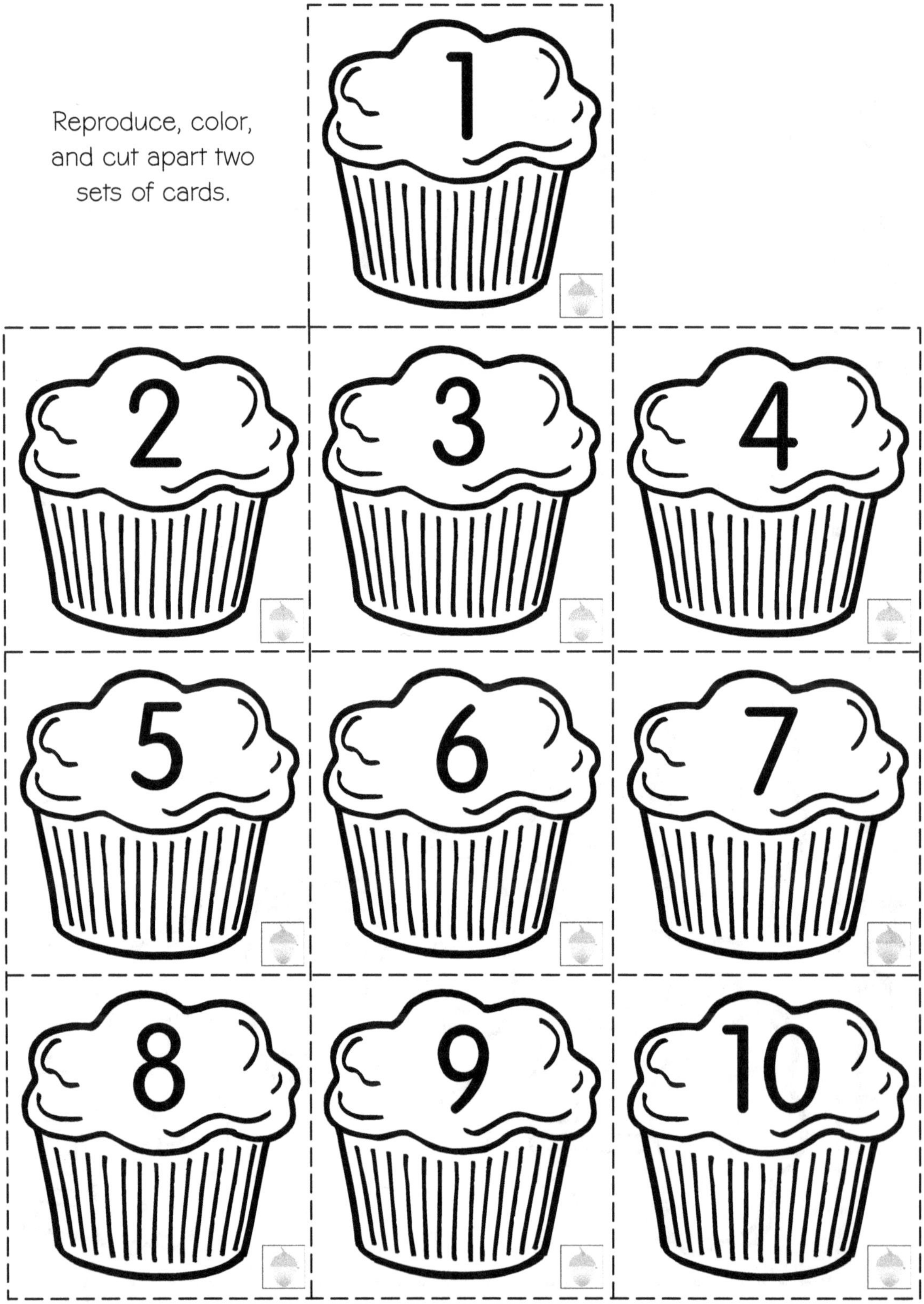

Ladybugs and Leaves
Matching Numerals and Number Words 1-10

Match Board

Assembly

Reproduce, cut out, and glue a set of oak tag handles (p. 192) to a standard-size file folder. Option: Decorate the handles with copies of the ladybugs shown on this page. Reproduce, color, and cut out the "Ladybug and Leaves" patterns. Glue the match boards to the inside of the folder. Glue the cover art to the front of the folder. Decorate the border around the match boards, then laminate. Reproduce, color, laminate, then cut apart three sets of game cards. Glue an envelope to the back of the folder for game card storage. Option: Reproduce, cut out, and glue the Match Board Playing Directions to the back of the folder.

Directions
Reproduce, cut out, and glue the direction strip to the game board.

Place each leaf on the ladybug with the matching number word.

Ladybugs and Leaves
Matching Numerals and Number Words 1-10

Match Board

Ladybugs and Leaves		
ten	one	six
four	nine	eight
seven	two	five
five	three	two

Ladybugs and Leaves
Matching Numerals and Number Words 1-10

Match Board

Ladybugs and Leaves

ten	four	five
three	one	one
three	six	seven
seven	eight	nine

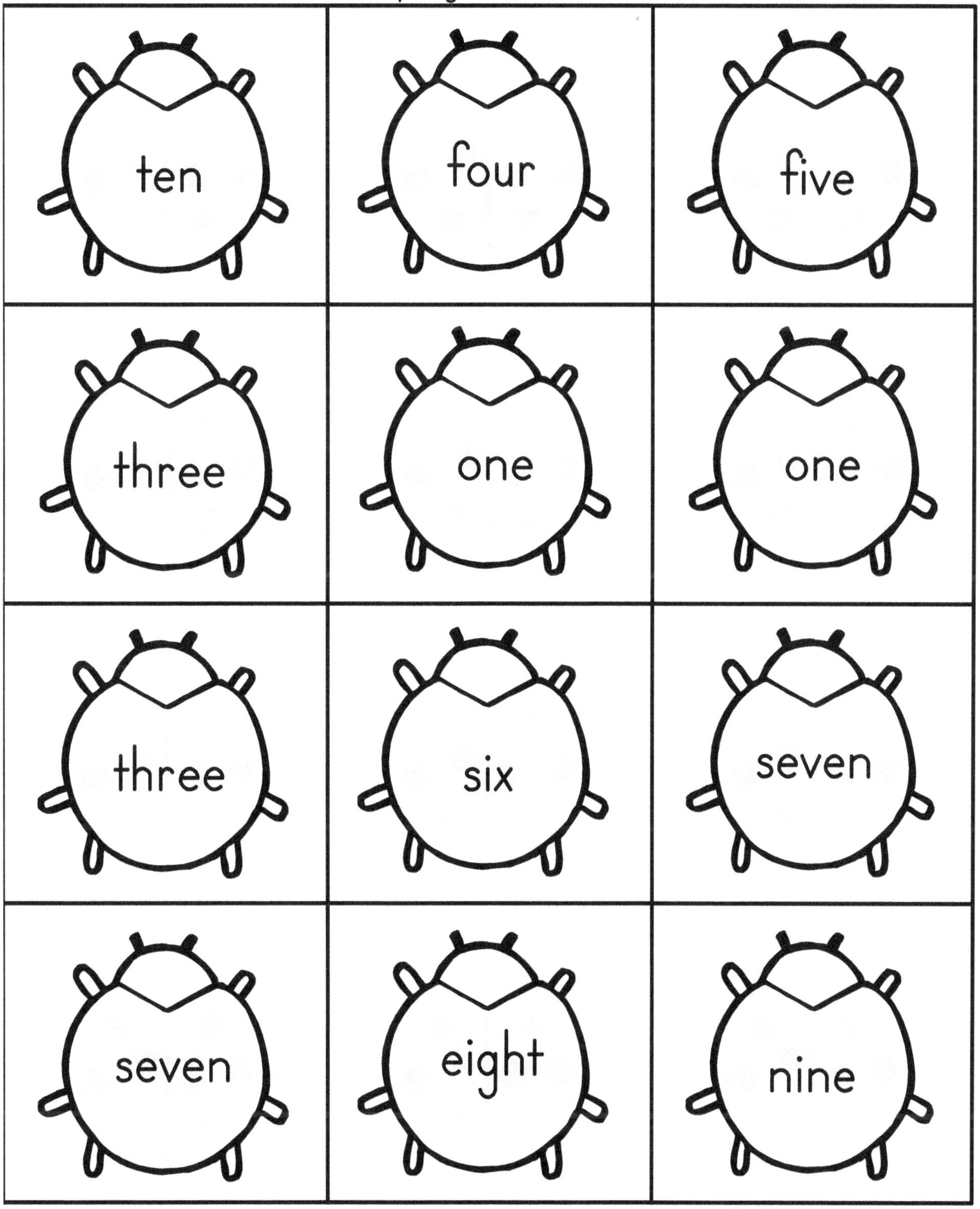

© 2014 Little Acorn Books™

Ladybugs and Leaves
Matching Numerals and Number Words 1-10

Match Board

Reproduce, color, and cut apart three sets of cards.

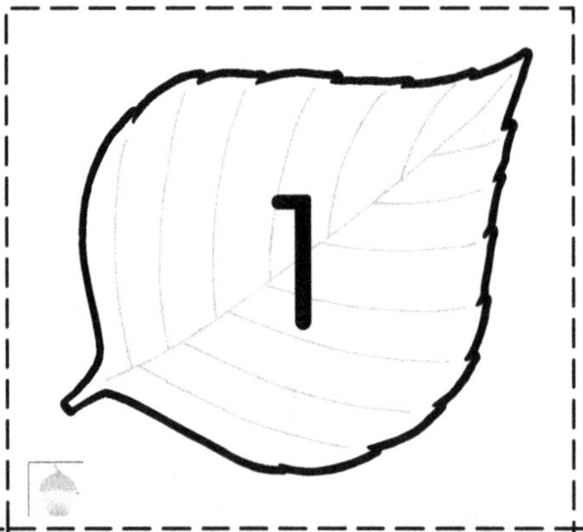

1

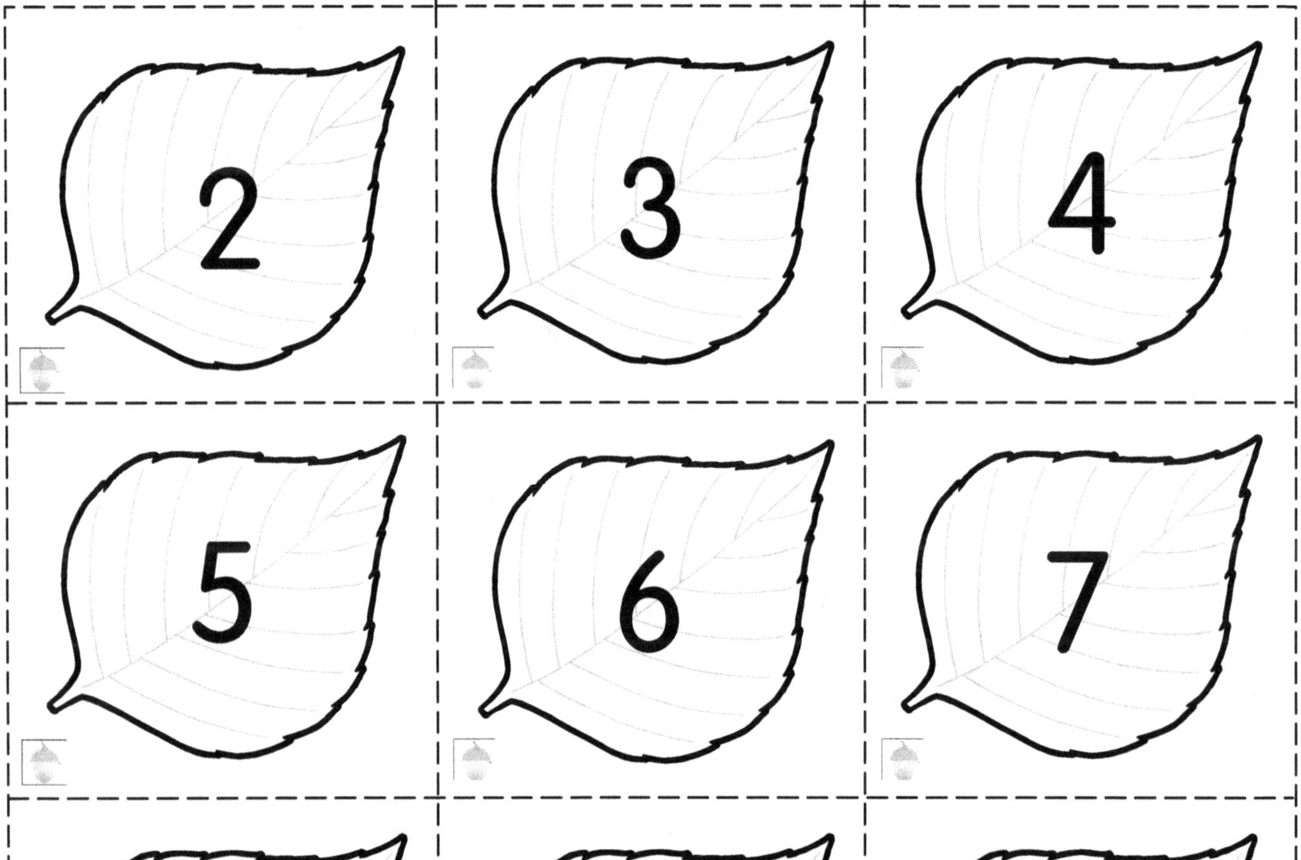

2　3　4

5　6　7

8　9　10

Ladybugs and Leaves Cover
Matching Numerals and Number Words 1-10

Match Board

Ladybugs and Leaves

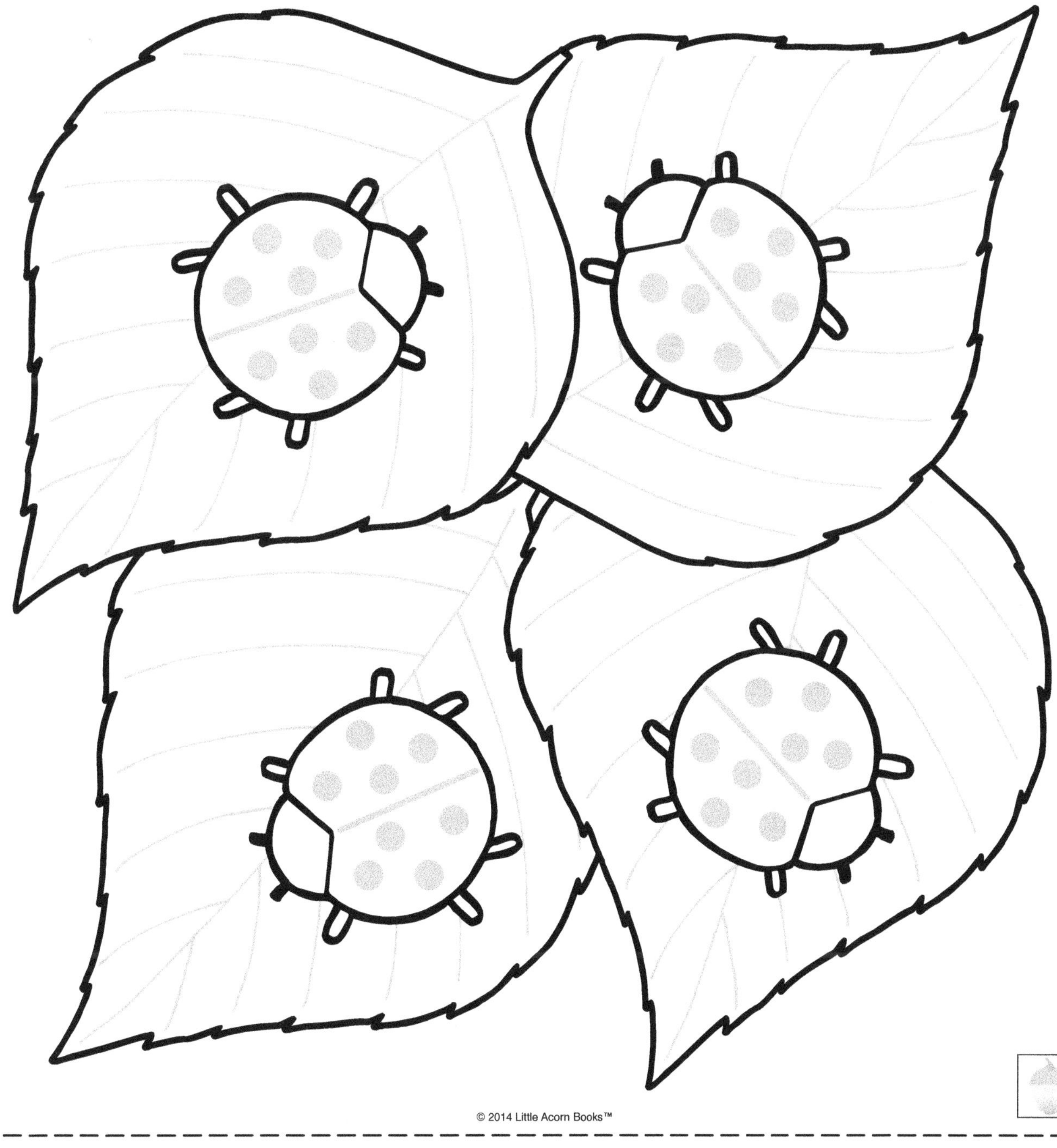

Rainbow Eggs Baskets
Matching Numerals and Number Sets 1-12

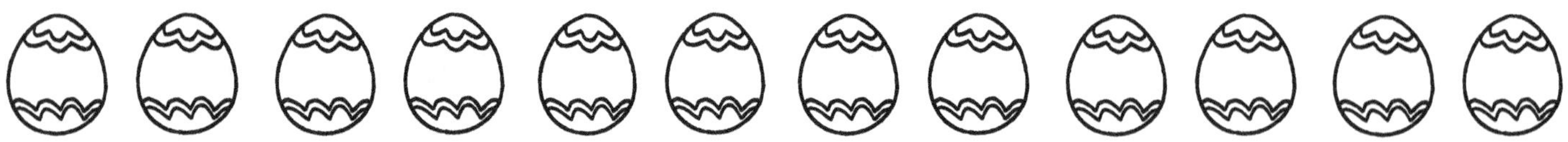

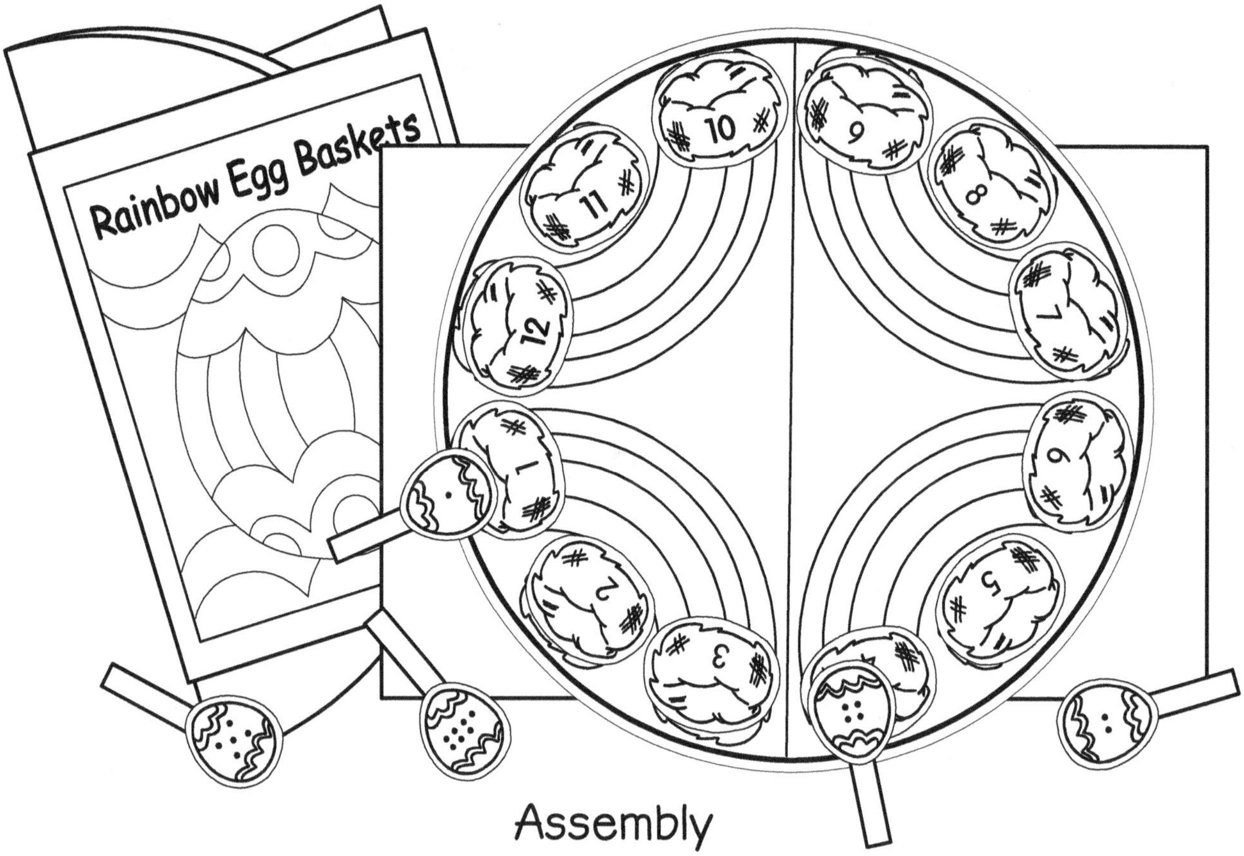

Clothespin Wheel

Assembly

Reproduce, cut out, and glue a set of oak tag handles (p. 192) to a standard-size file folder. Option: Decorate the handles with copies of the eggs shown on this page. Reproduce, color, and cut out the "Rainbow Eggs Basket" patterns. Glue the cover art to the front of the folder. Staple a resealable plastic bag to the back of the folder for clothespin storage. Glue the polka-dot eggs on clothespins. Meeting along the straight edges, glue the basket wheel quarters on a large poster board circle. Glue the wheel inside the folder, then glue the numeral nests on the wheel. Option: Reproduce, cut out, and glue the Clothespin Wheel directions to the back of the folder.

Directions
Reproduce, cut out, and glue the direction strip to the game board.

Pin each egg to a matching nest.

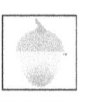

Rainbow Eggs Baskets Cover
Matching Numerals and Number Sets 1-12

Clothespin Wheel

Rainbow Egg Baskets

Rainbow Eggs Baskets

Clothespin Wheel

Matching Numerals and Number Sets 1-12

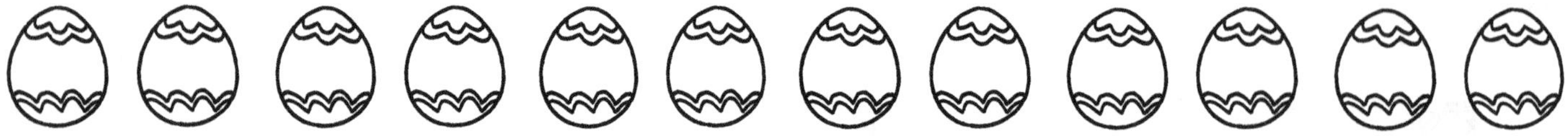

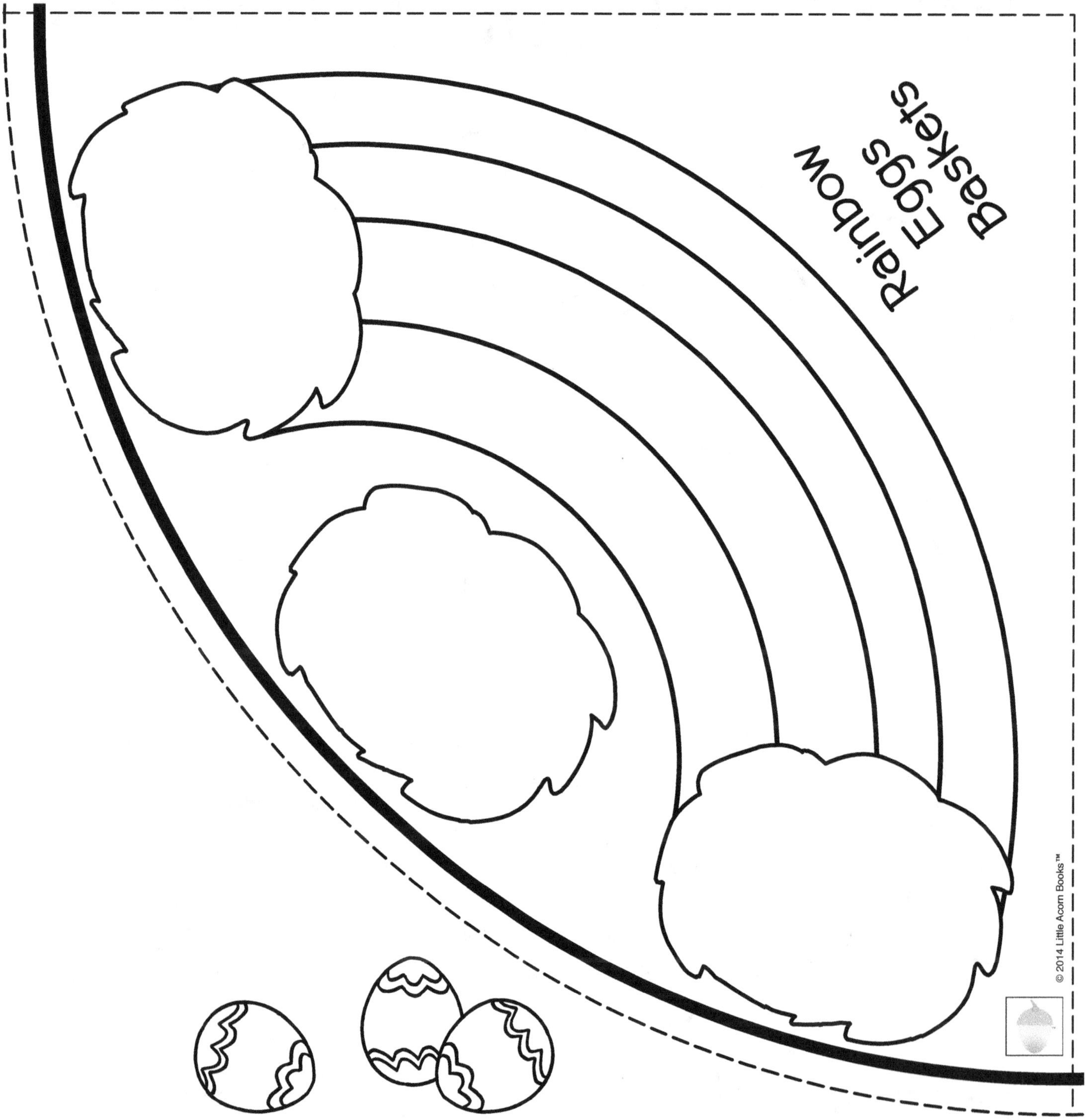

Reproduce, color, and cut out four wheel quarters.
Meeting along the straight edges, glue the quarters on a large poster board circle.

LAB20145P • 35 FOLDERGAMES FOR NUMBERS • 978-1-937257-53-8 • © 2014 Little Acorn Books™

Rainbow Eggs Baskets
Matching Numerals and Number Sets 1-12

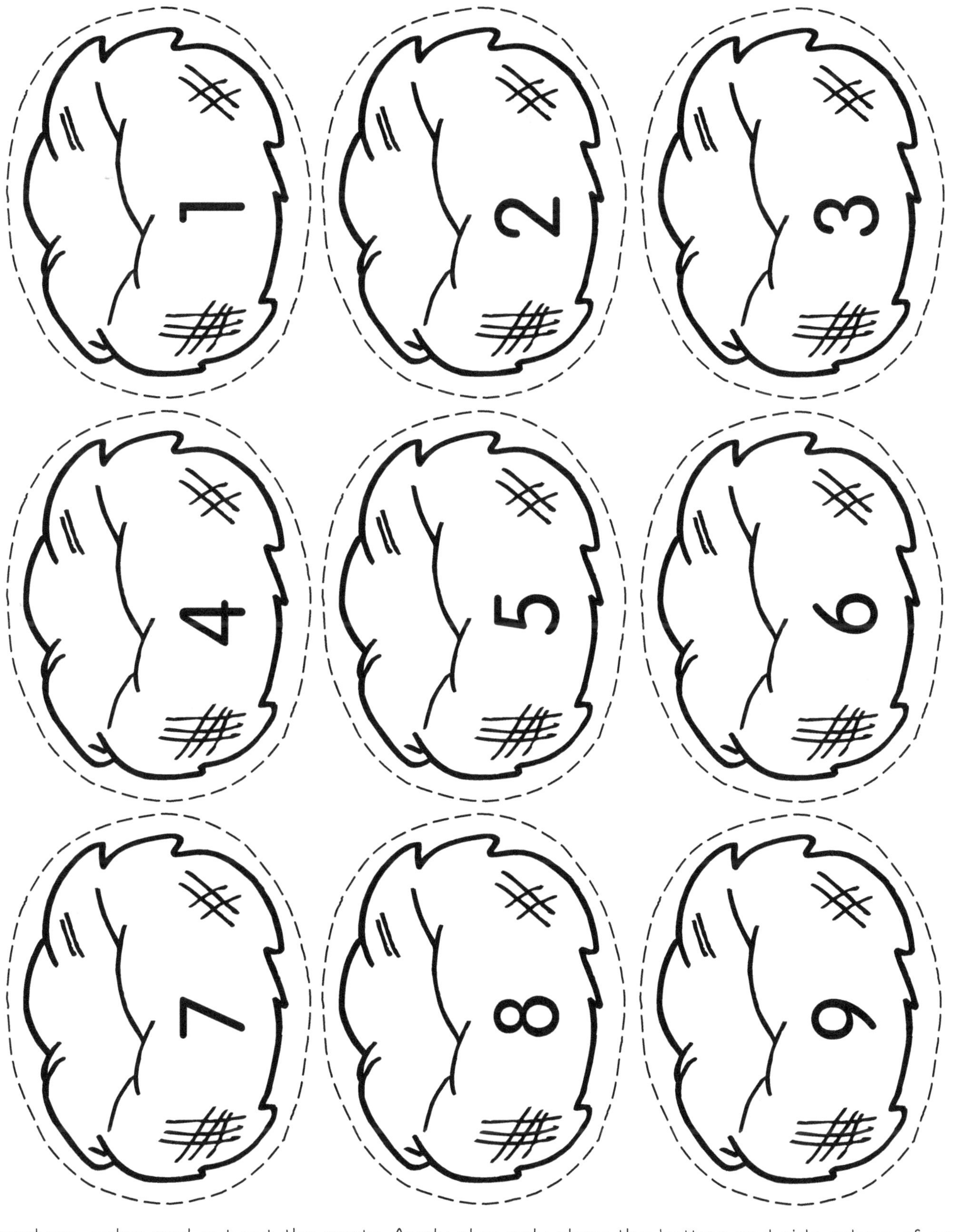

Clothespin Wheel

Reproduce, color, and cut out, the nests. Apply glue only along the bottom and side edges of each nest. Attach each nest to the inside of a folder.

Rainbow Eggs Baskets
Matching Numerals and Number Sets 1-12

Clothespin Wheel

Reproduce, color, cut out, and glue the eggs on clothespins
Include the eggs below for advanced skills practice.

Reproduce, color, and cut out the nests. Apply glue only along the bottom and side edges of each nest. Attach each nest to the inside of a folder.

Firefly Lights
Matching Numerals and Number Sets 1-12

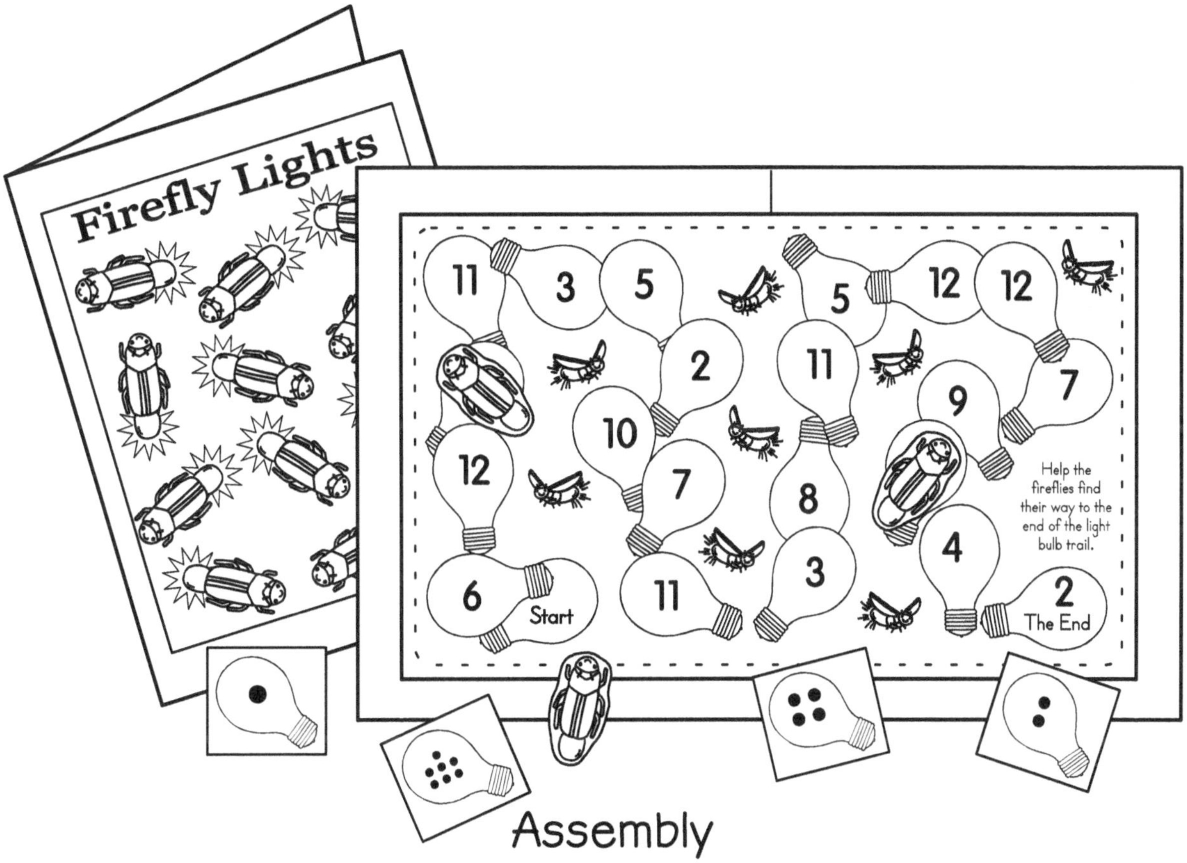

Trail Game

Assembly

Reproduce, cut out, and glue a set of oak tag handles (p. 192) to a standard-size file folder. Option: Decorate each handle with a copy of the light bulbs on this page. Reproduce, color, and cut out the "Firefly Lights" patterns. Matching in the center, glue the game board patterns to the inside of the folder. Glue the cover art to the front of the folder. Decorate the border around the game board, then laminate. Reproduce, color, laminate, then cut out the pawns and two sets of game cards. Glue an envelope to the back of the folder for pawn and game card storage. Option: Reproduce, cut out, and glue the Trail Game Playing Directions to the back of the folder.

Pawns
Reproduce, color, and cut out.

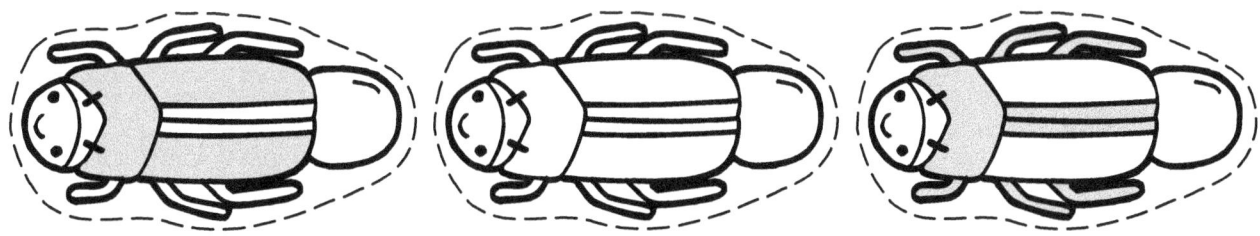

Firefly Lights
Matching Numerals and Number Sets 1-12

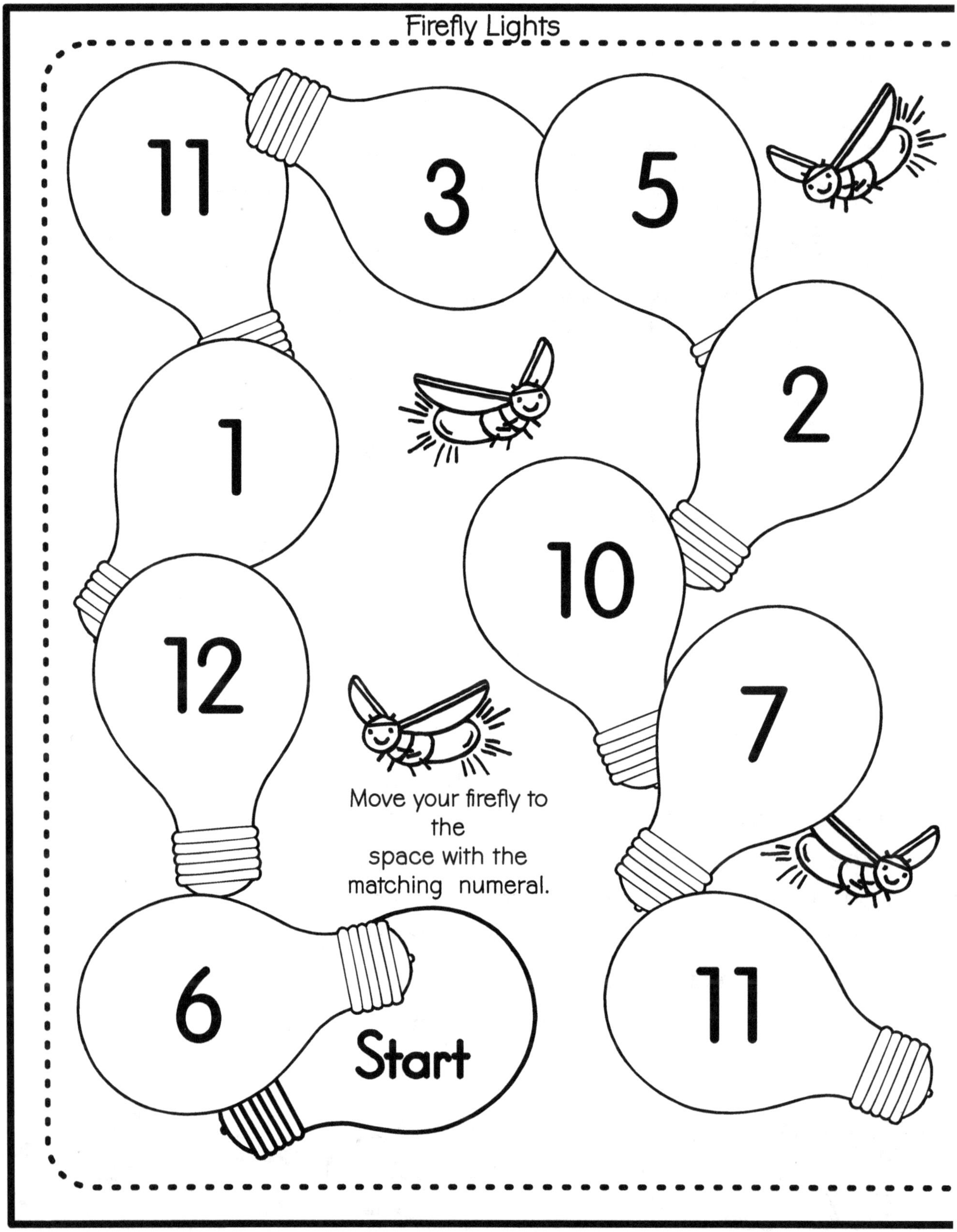

Trail Game

Firefly Lights

11 3 5

1 2

12 10 7

6 Start 11

Move your firefly to the space with the matching numeral.

Firefly Lights

Trail Game

Matching Numerals and Number Sets 1-12

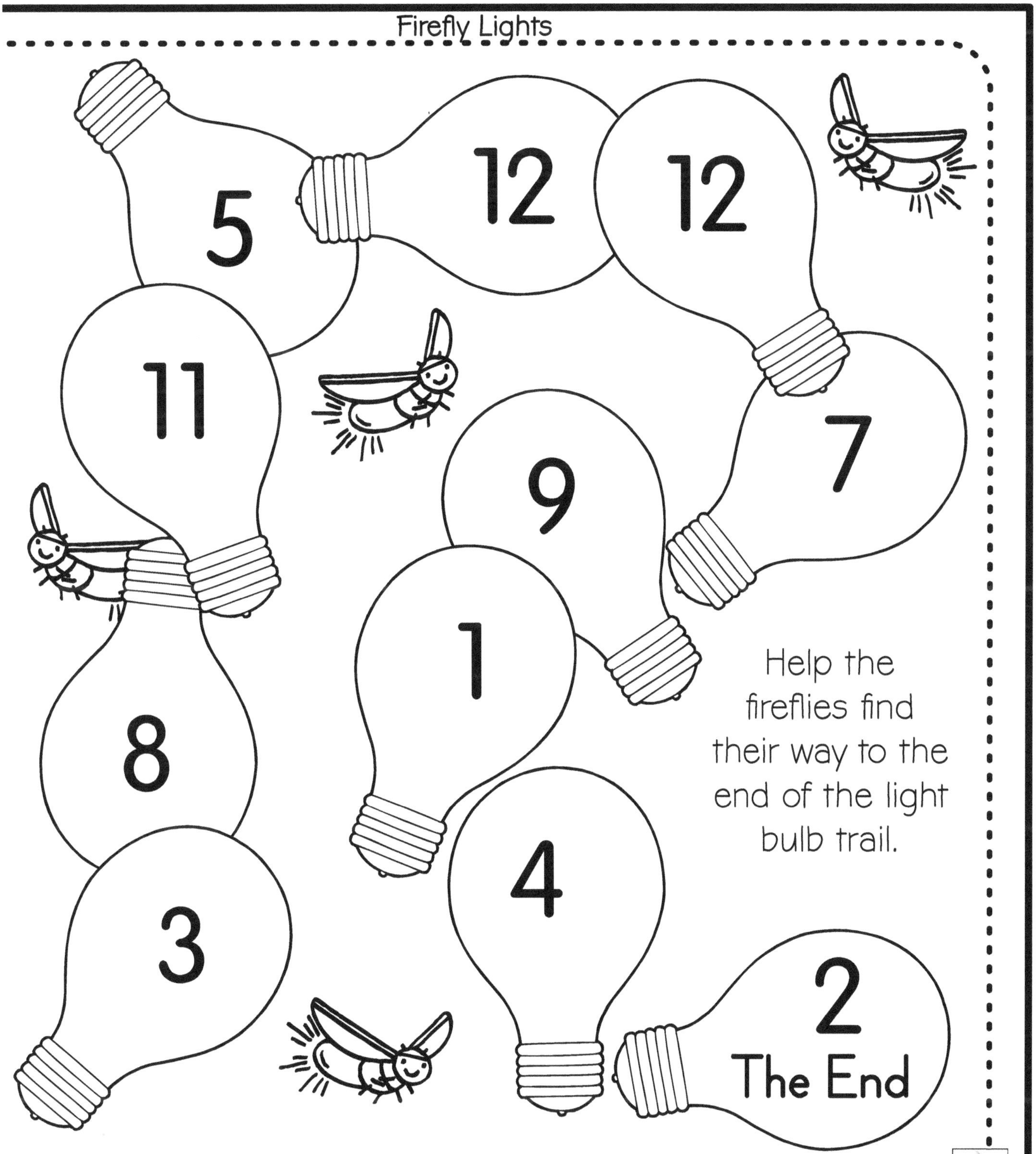
Firefly Lights
5
12
12
11
7
9
1
8
Help the fireflies find their way to the end of the light bulb trail.
4
3
2
The End
© 2014 Little Acorn Books™

Firefly Lights

Trail Game

Matching Numerals and Number Sets 1-12

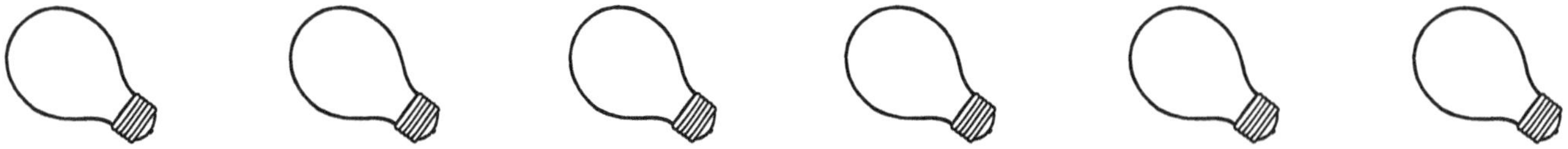

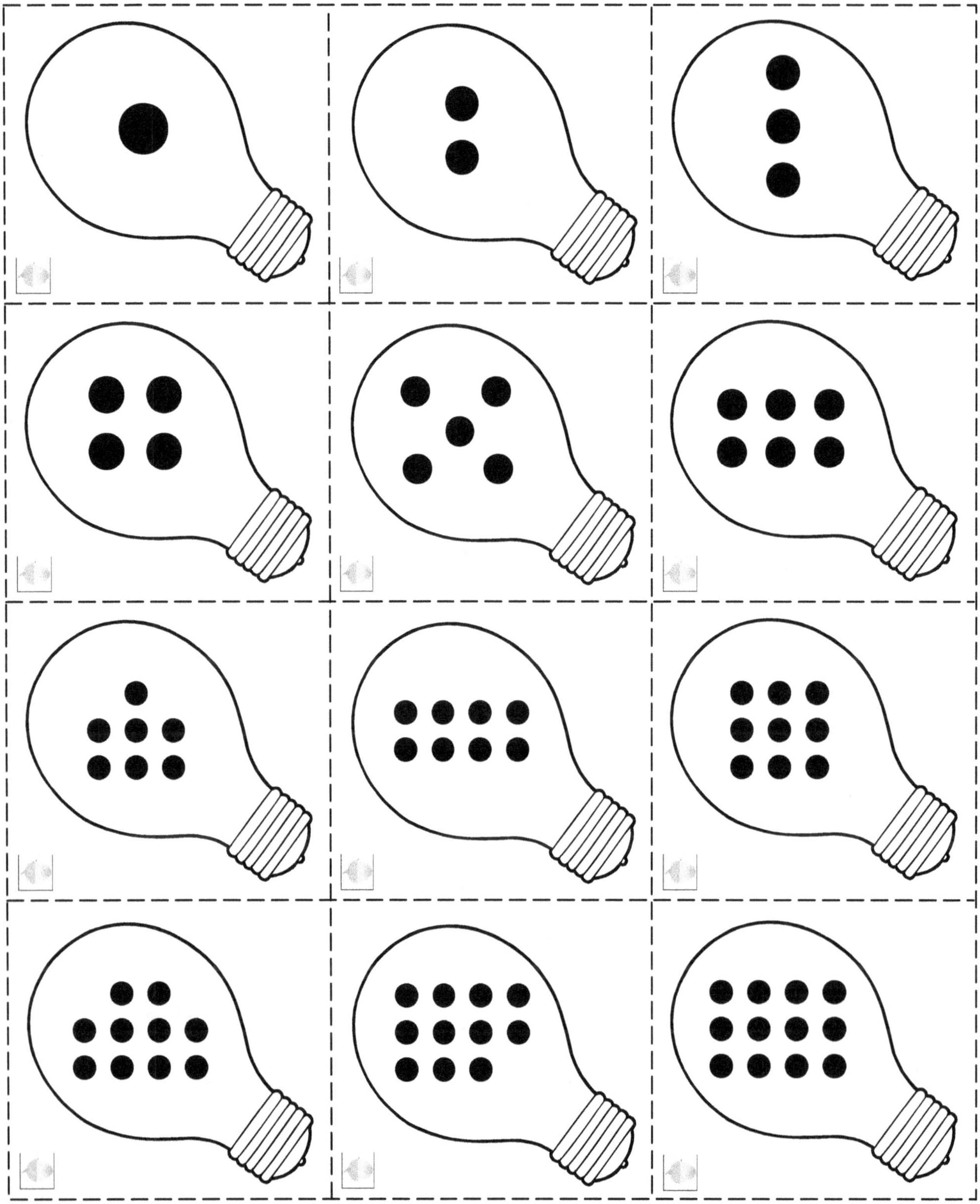

Reproduce, color, and cut apart two sets of cards.

LAB20145P • 35 FOLDERGAMES FOR NUMBERS • 978-1-937257-53-8 • © 2014 Little Acorn Books™

Trail Game

Firefly Lights Cover
Matching Numerals and Number Sets 1-12

Firefly Lights

Polka-Dot Ponies Lotto
Matching Numerals and Number Sets 1-12

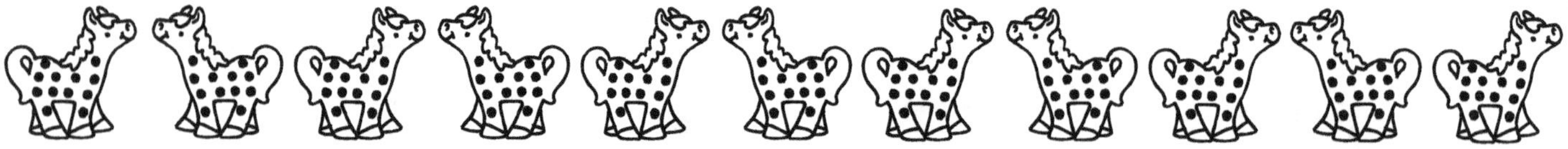

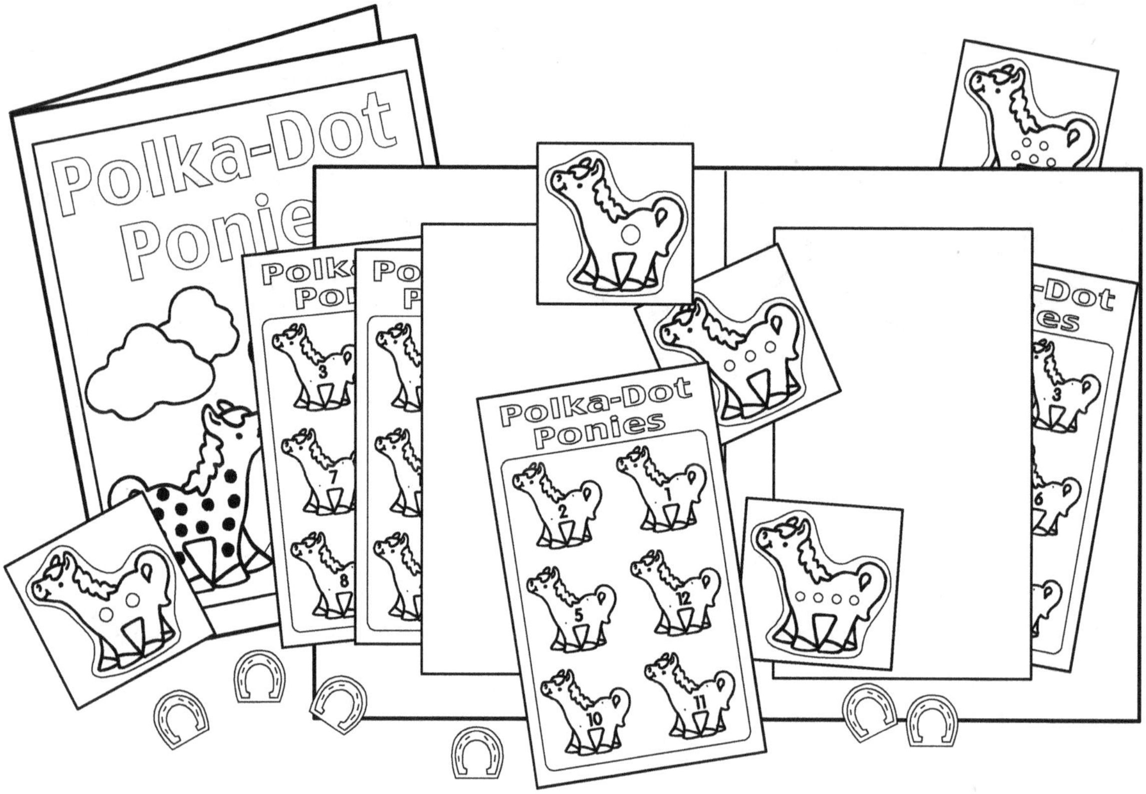

Lotto

Assembly

Reproduce, cut out, and glue a set of oak tag handles (p. 192) to a standard-size file folder. Option: Decorate each handle with a copy of the ponies shown above. Reproduce, color, and cut out the "Polka-Dot Ponies Lotto" patterns. Glue the cover art to the front of the folder. Mount Lotto boards on colored construction paper, then laminate. Glue two envelopes to the inside of the folder for Lotto board, token, and card storage. Option: Reproduce, cut out, and glue the Lotto Playing Directions to the back of the folder. Note: Wipe-off markers or crayons can also be used to color in matches.

Tokens
Reproduce, color, and cut out tokens for players to place on Lotto boards.

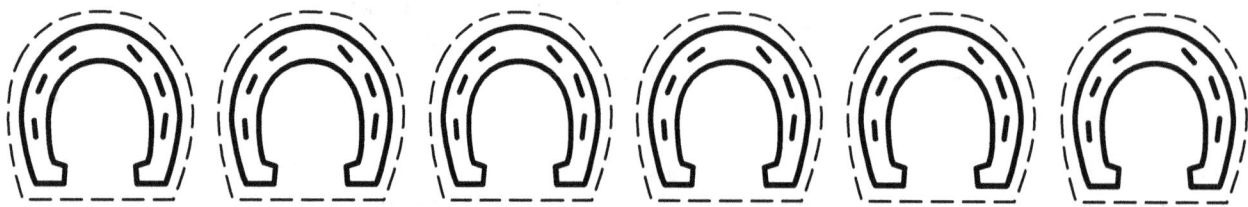

Lotto

Polka-Dot Ponies Lotto Cover
Matching Numerals and Number Sets 1-12

Polka-Dot Ponies

Polka-Dot Ponies Lotto
Matching Numerals and Number Sets 1-12

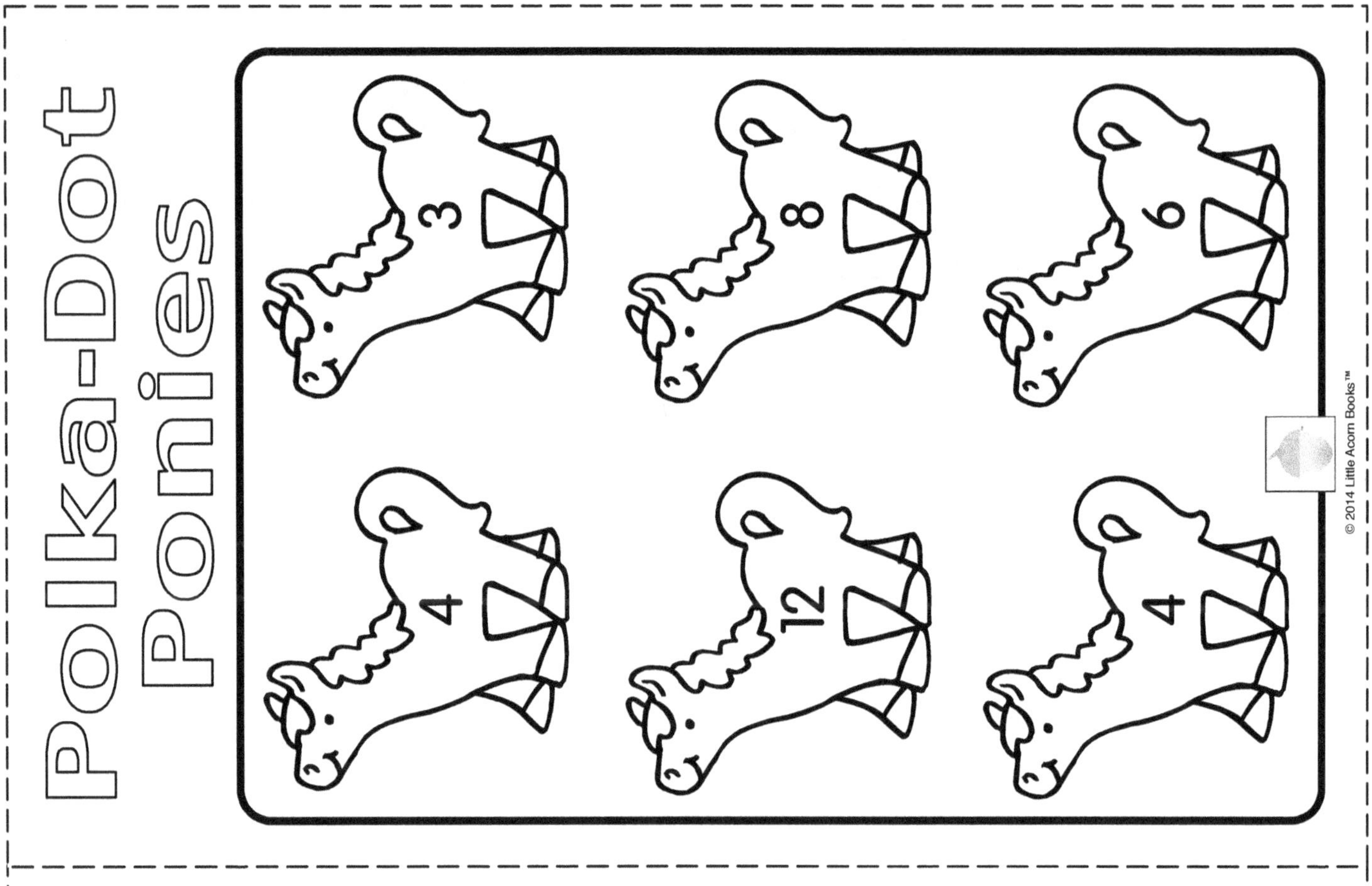

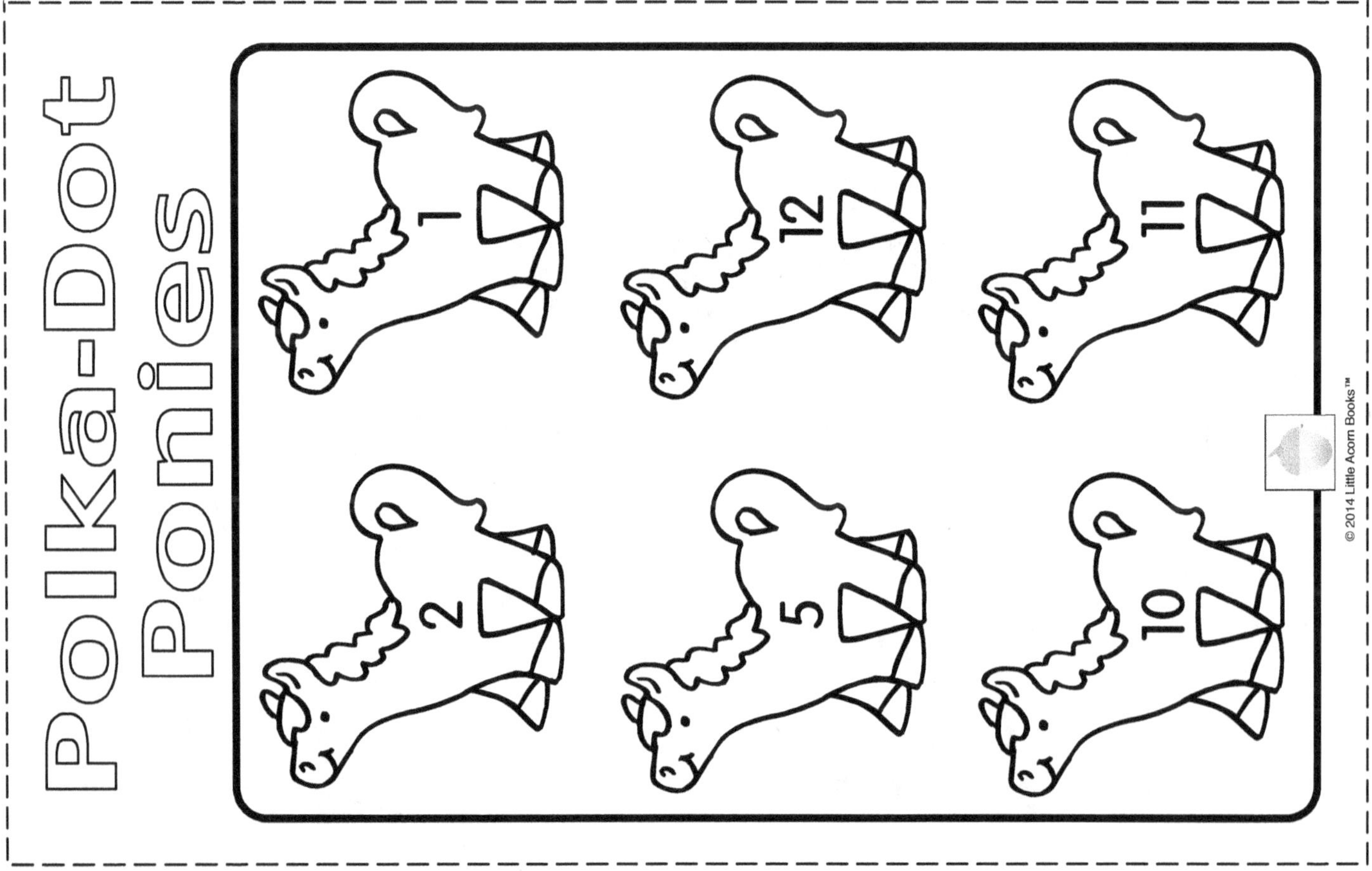

Lotto

Polka-Dot Ponies

3, 8, 9
4, 12, 4

Polka-Dot Ponies

1, 12, 11
2, 5, 10

Polka Dot-Ponies Lotto
Matching Numerals and Number Sets 1-12

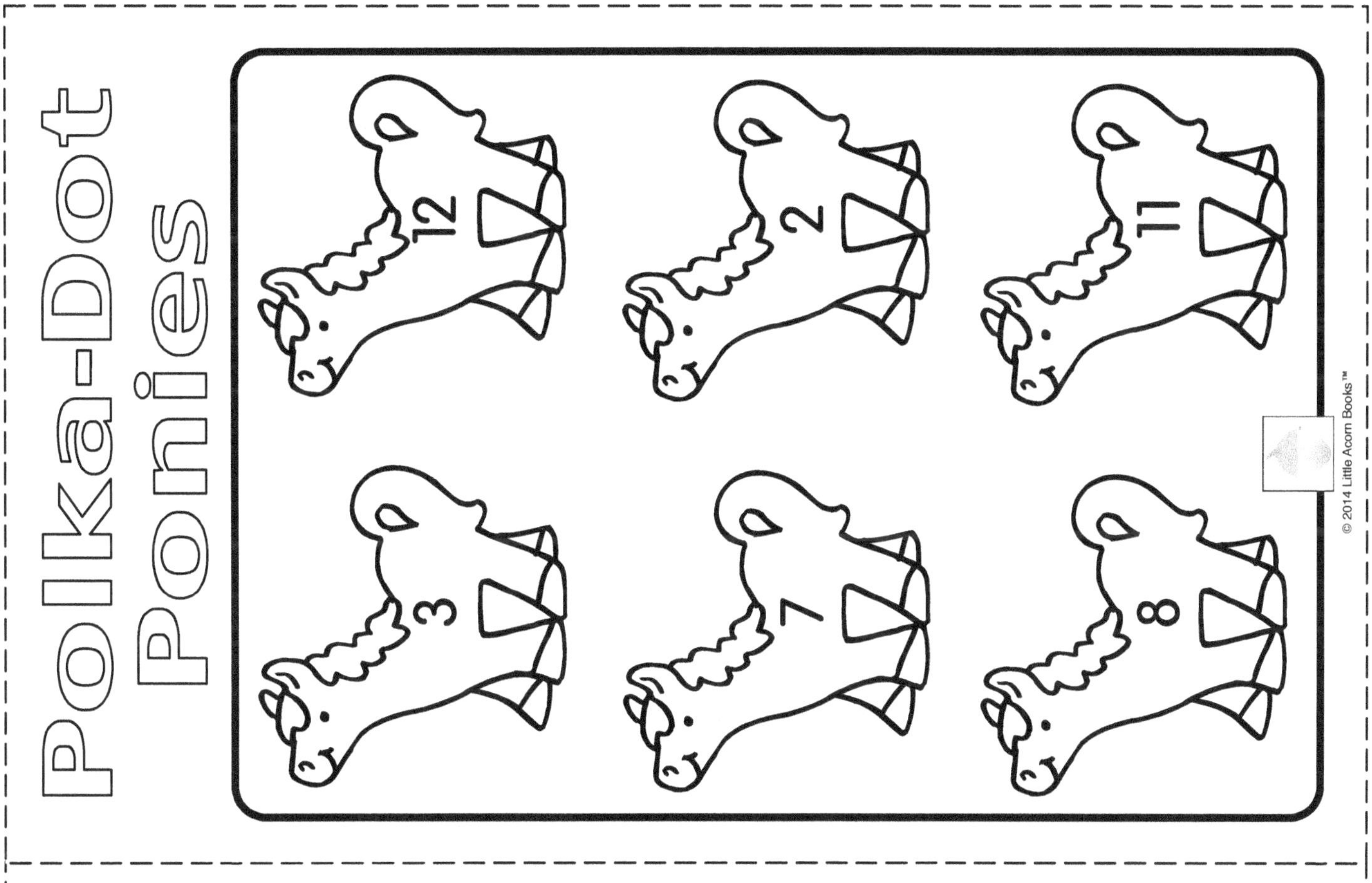

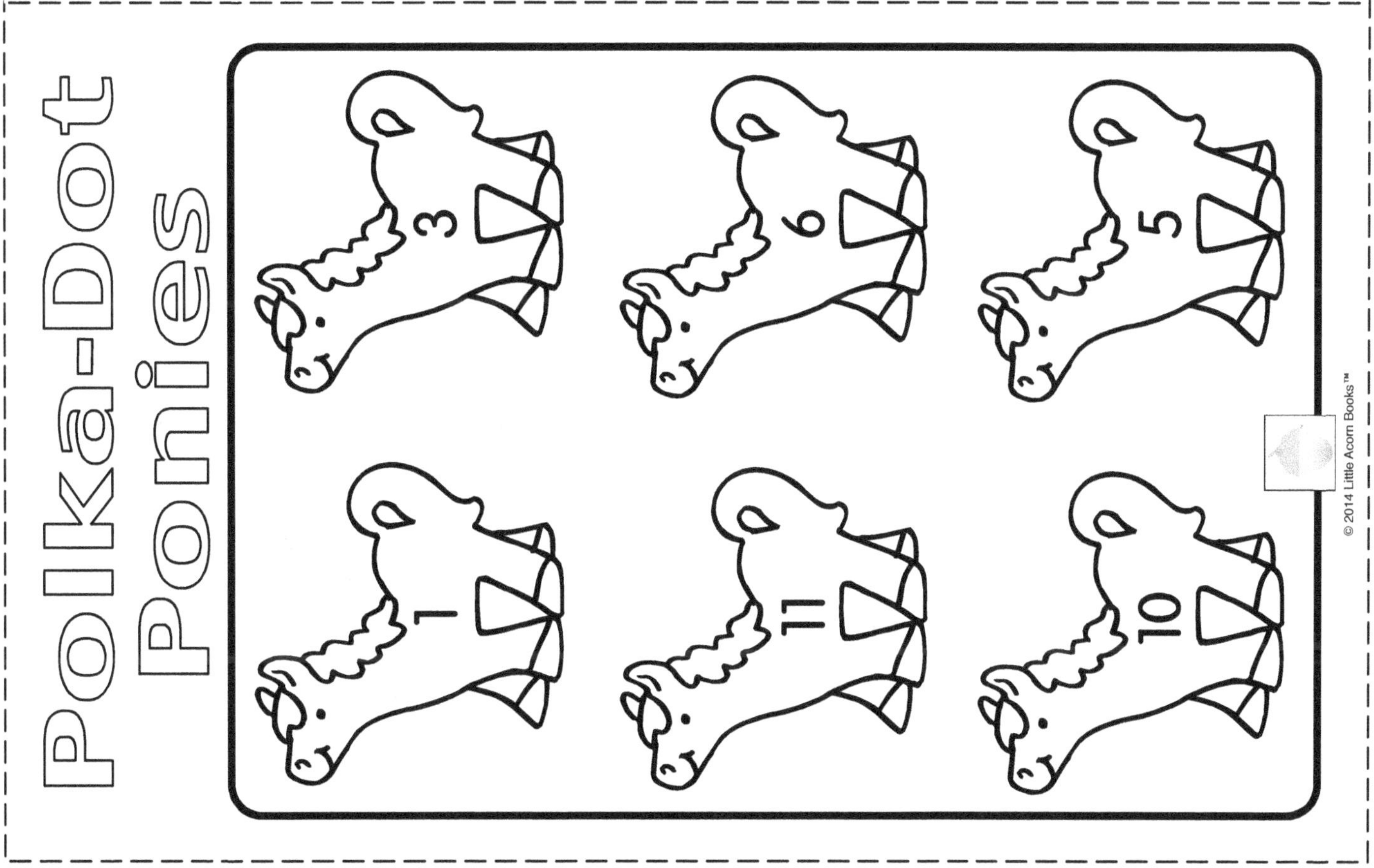

Lotto

Polka-Dot Ponies

12, 2, 11
3, 7, 8

Polka-Dot Ponies

3, 6, 5
1, 11, 10

Polka-Dot Ponies Lotto
Matching Numerals and Number Sets 1-12

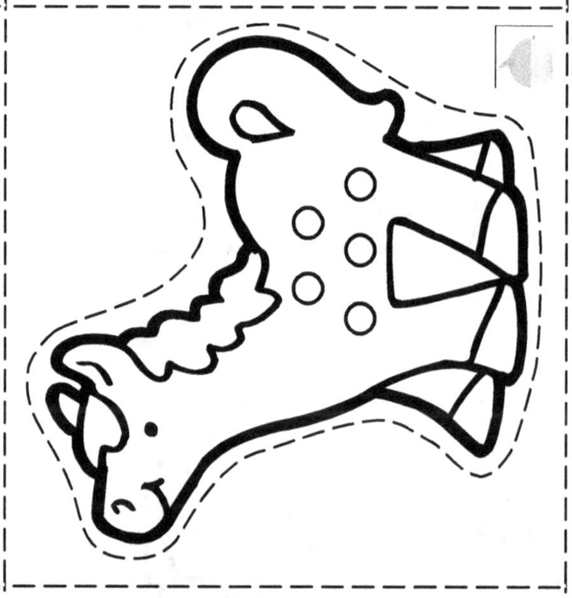

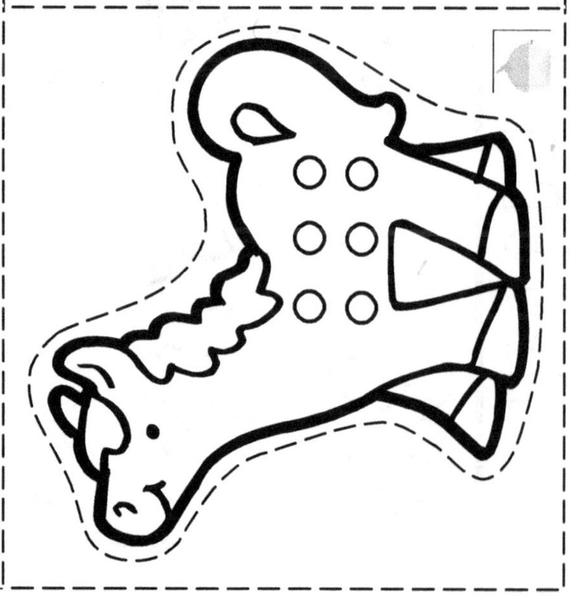

Lotto

128 LAB20145P • 35 FOLDERGAMES FOR NUMBERS • 978-1-937257-53-8 • © 2014 Little Acorn Books™

Polka-Dot Ponies Lotto
Matching Numerals and Number Sets 1-12

Lotto

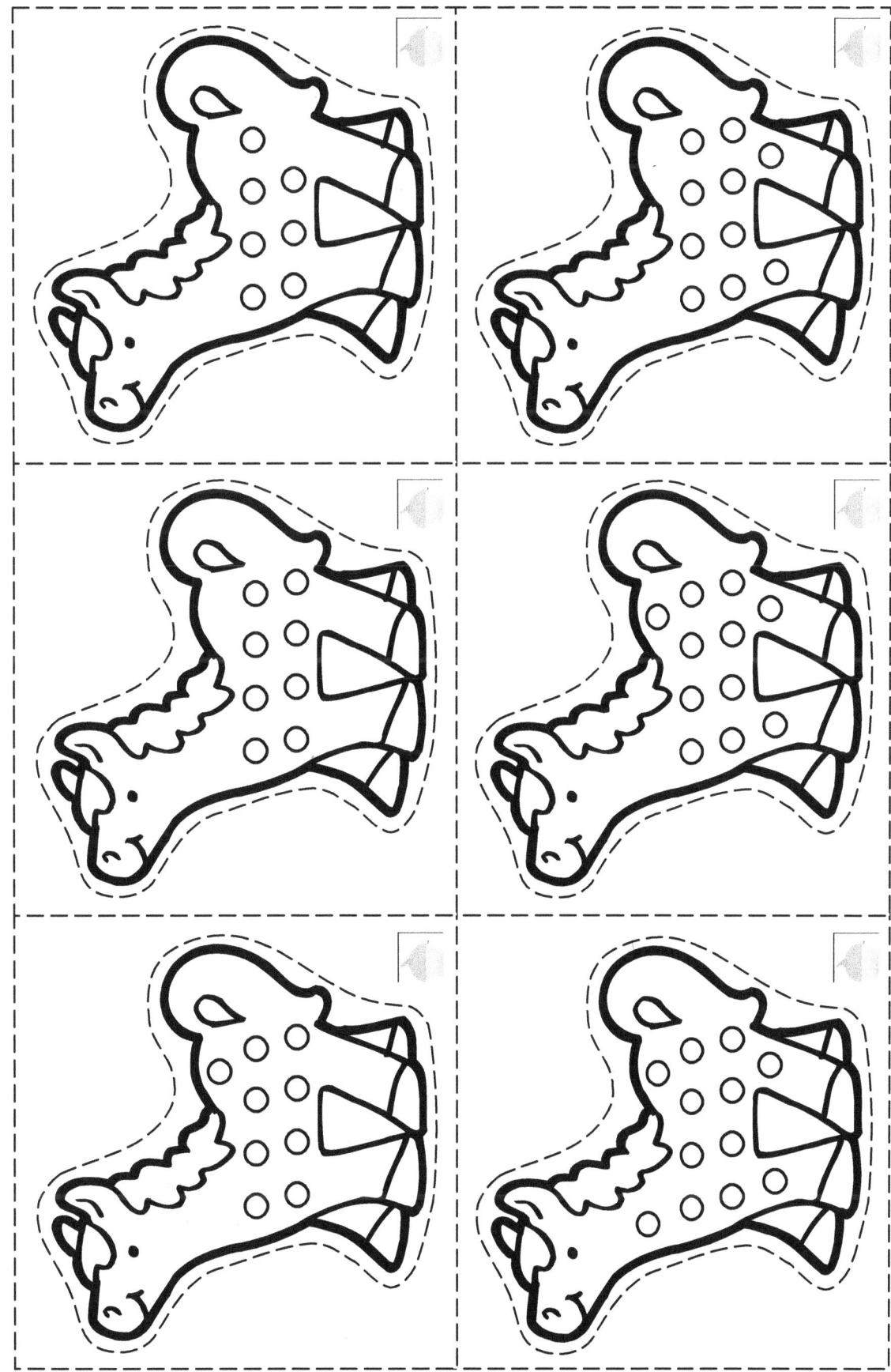

LAB20145P • 35 FOLDERGAMES FOR NUMBERS • 978-1-937257-53-8 • © 2014 Little Acorn Books™

Carnival Clown Bingo
Matching Number Words and Numerals 1-12

Bingo

Assembly

Reproduce, cut out, and glue a set of oak tag handles (p. 192) to a standard-size file folder. Option: Decorate each handle with a copy of the clown hats shown above. Reproduce, color, and cut out the "Carnival Clown Bingo" patterns. Mount Bingo boards on colored construction paper, then laminate. Glue the cover art to the front of the folder. Glue two envelopes to the inside of the folder for Bingo board, token, and card storage. Option: Reproduce, cut out, and glue the Bingo Playing Directions to the back of the folder. Note: Wipe-off markers or crayons can be also be used to color in matches.

Tokens
Reproduce, color, and cut out tokens for players to place on Bingo Boards.

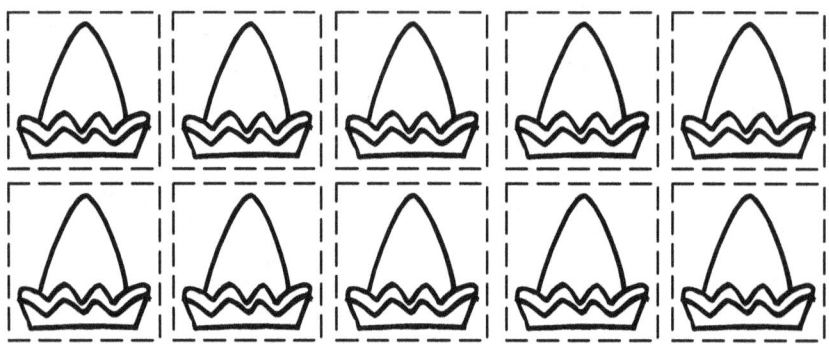

Carnival Clown Bingo Cover
Matching Number Words and Numerals 1-12

Carnival Clown Bingo

Carnival Clown Bingo
Matching Number Words and Numerals 1-12

C	L	O	W	N
eight	twelve	ten	nine	six
nine	six	five	seven	four
seven	four	FREE SPACE	five	one
one	ten	eight	twelve	nine
three	seven	two	two	eight

C	L	O	W	N
eight	twelve	four	seven	seven
two	one	six	five	nine
ten	seven	FREE SPACE	three	four
five	eleven	ten	nine	two
one	six	nine	eight	twelve

Carnival Clown Bingo
Matching Number Words and Numerals 1-12

Bingo

C	L	O	W	N
eleven	nine	ten	eleven	eight
four	seven	five	nine	two
one	five	FREE SPACE	eleven	two
nine	eleven	twelve	ten	seven
eight	nine	seven	one	three

C	L	O	W	N
five	five	four	two	ten
six	twelve	three	nine	eight
four	six	FREE SPACE	twelve	nine
five	eight	seven	eleven	six
eight	two	ten	seven	one

Carnival Clown Bingo
Matching Number Words and Numerals 1-12

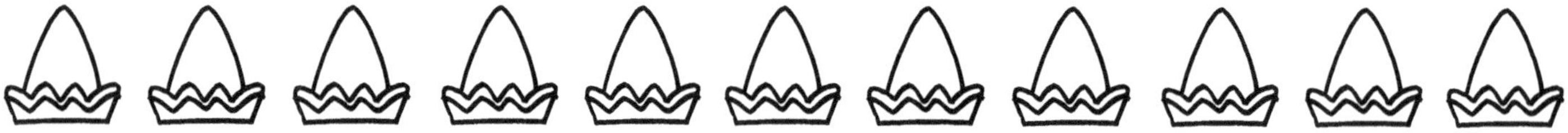

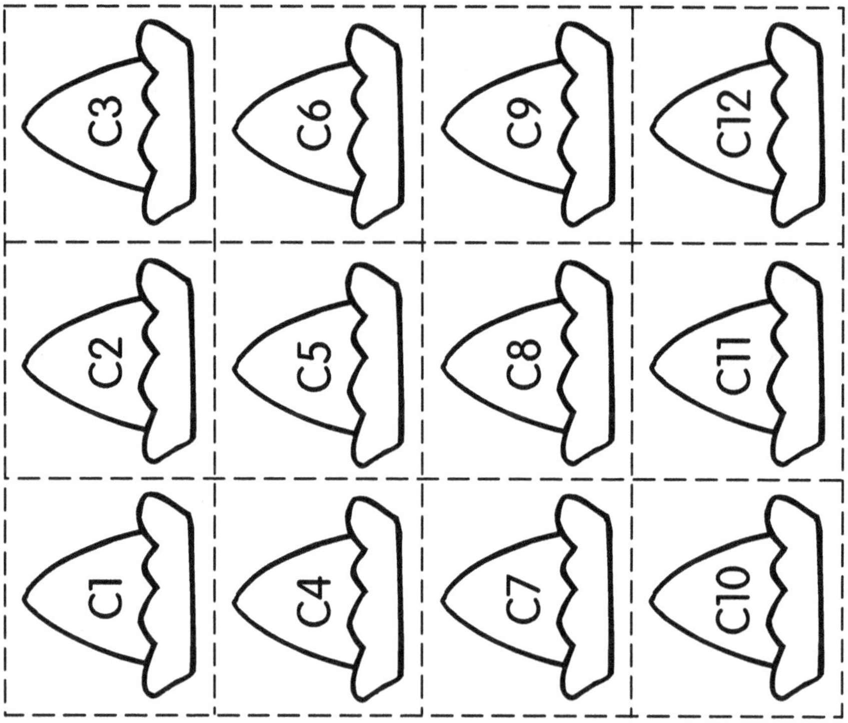

Bingo

C3	C6	C9	C12
C2	C5	C8	C11
C1	C4	C7	C10

L1	L2	L3	O1	O2	O3
L4	L5	L6	O4	O5	O6
L7	L8	L9	O7	O8	O9
L10	L11	L12	O10	O11	O12

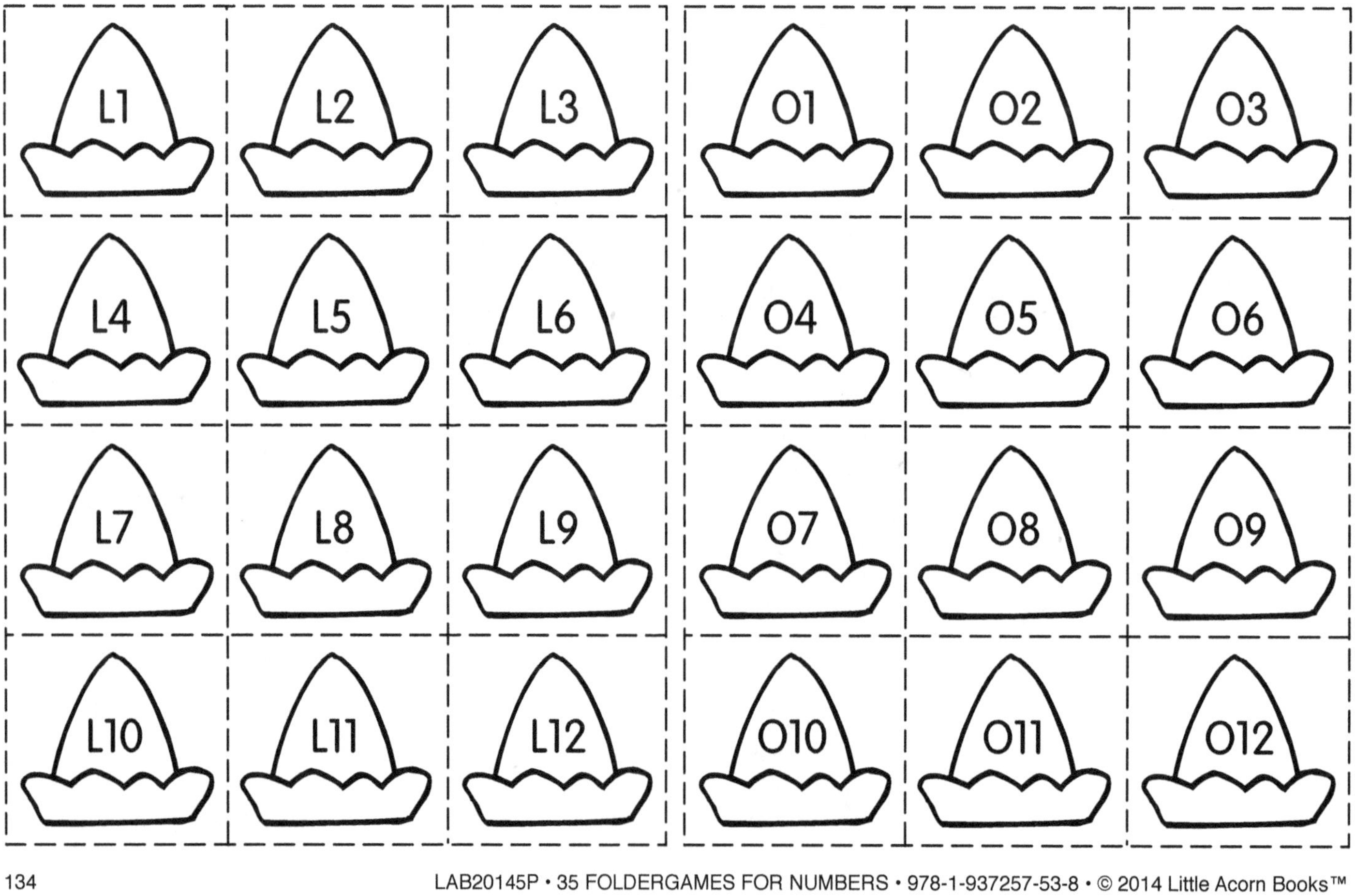

Carnival Clown Bingo
Matching Number Words and Numerals 1-12

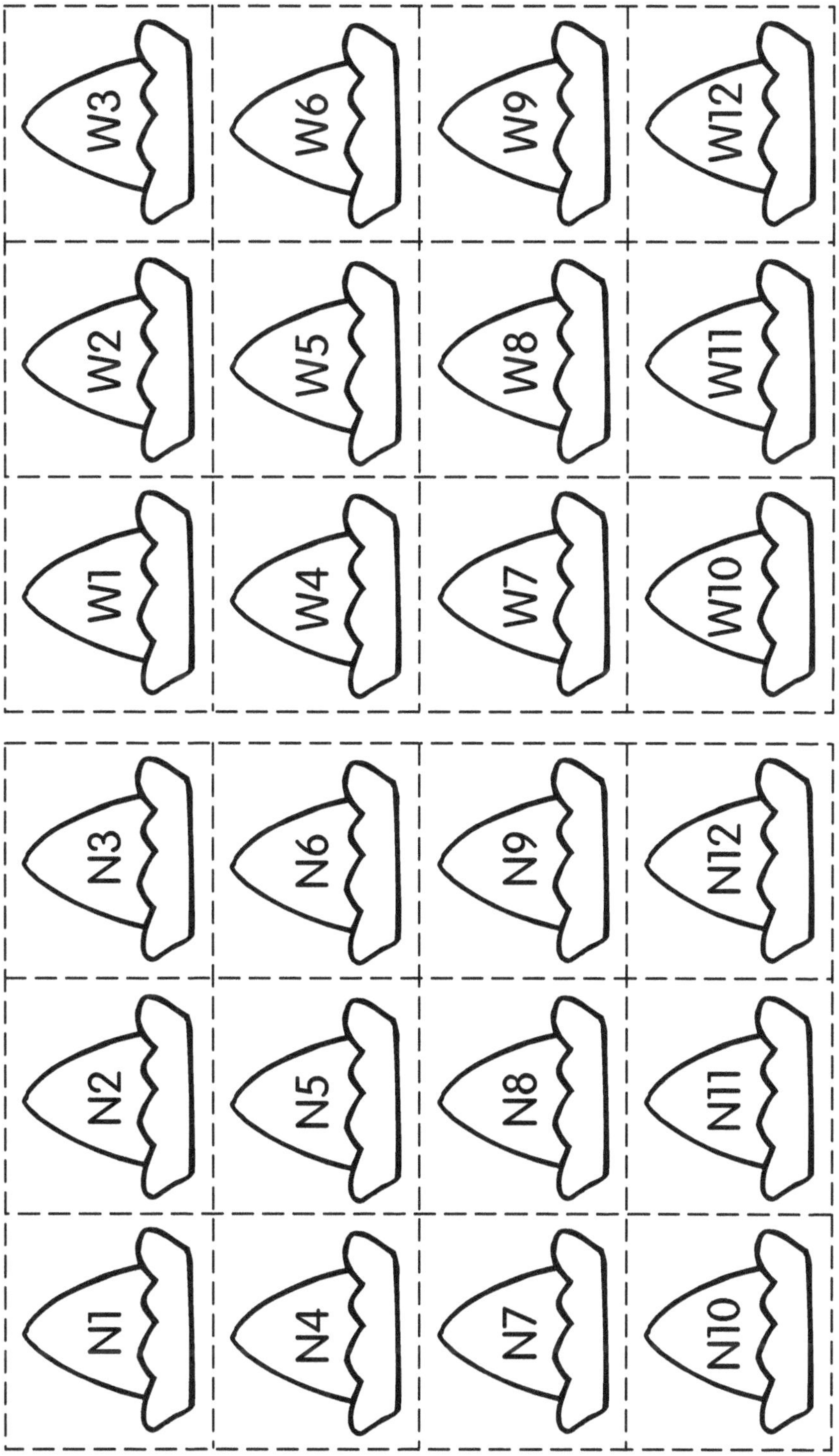

Bingo

W3	W6	W9	W12
W2	W5	W8	W11
W1	W4	W7	W10
N3	N6	N9	N12
N2	N5	N8	N11
N1	N4	N7	N10

Turtle Dominoes
Matching Number Words and Numerals 1-12

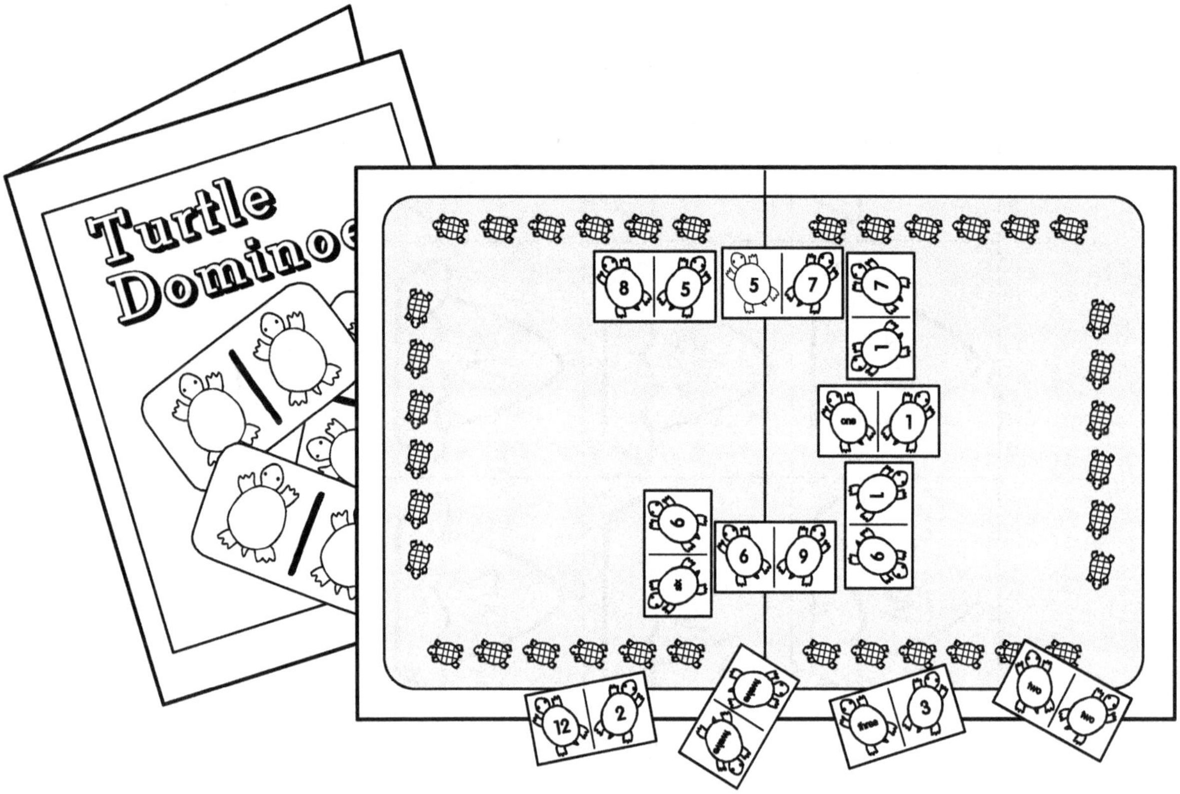

Dominoes

Assembly

Reproduce, cut out, and glue a set of oak tag handles (p. 192) to a standard-size file folder. Option: Decorate the handles with copies of the turtles shown on this page. Reproduce, color, and cut out the "Turtle Dominoes" patterns. Glue the cover art to the front of the folder. Decorate the inside of the folder, then laminate. Reproduce, color, laminate, then cut apart the dominoes. Glue an envelope to the back of the folder to store dominoes. Option: Reproduce, cut out, and glue the Dominoes Playing Directions to the back of the folder.

Turtle Dominoes Cover
Matching Number Words and Numerals 1-12

Dominoes

Turtle Dominoes

Turtle Dominoes
Matching Number Words and Numerals 1-12

Dominoes

one	1	seven	7	1	7
two	2	eight	8	6	9
three	3	nine	9	12	2
four	4	ten	10	8	5
five	5	eleven	11	3	11
six	6	twelve	12	4	10

Turtle Dominoes
Matching Number Words and Numerals 1-12

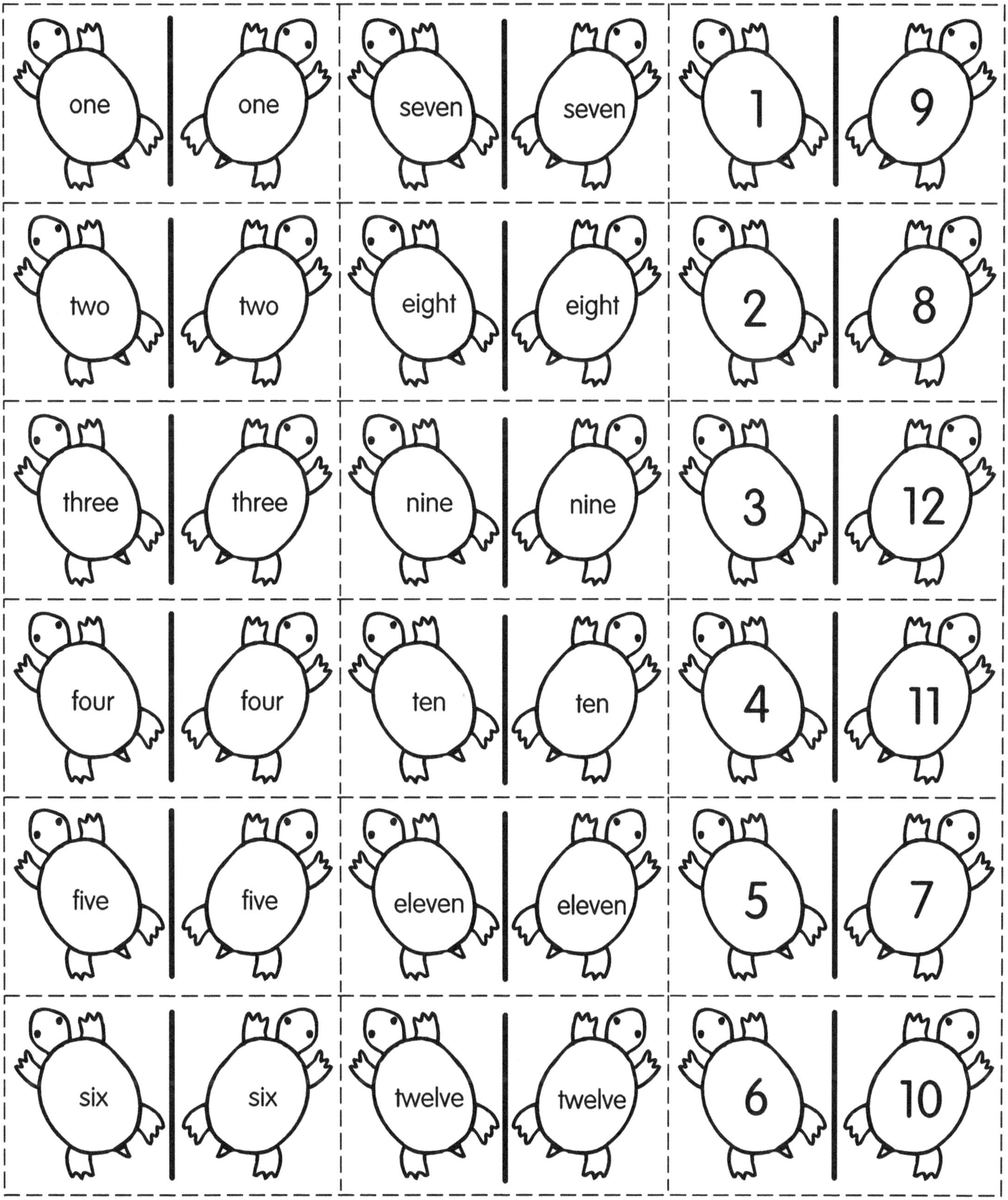

Dominoes

one	one	seven	seven	1	9
two	two	eight	eight	2	8
three	three	nine	nine	3	12
four	four	ten	ten	4	11
five	five	eleven	eleven	5	7
six	six	twelve	twelve	6	10

Zebra Zippers
Sequencing Number Sets 1-12

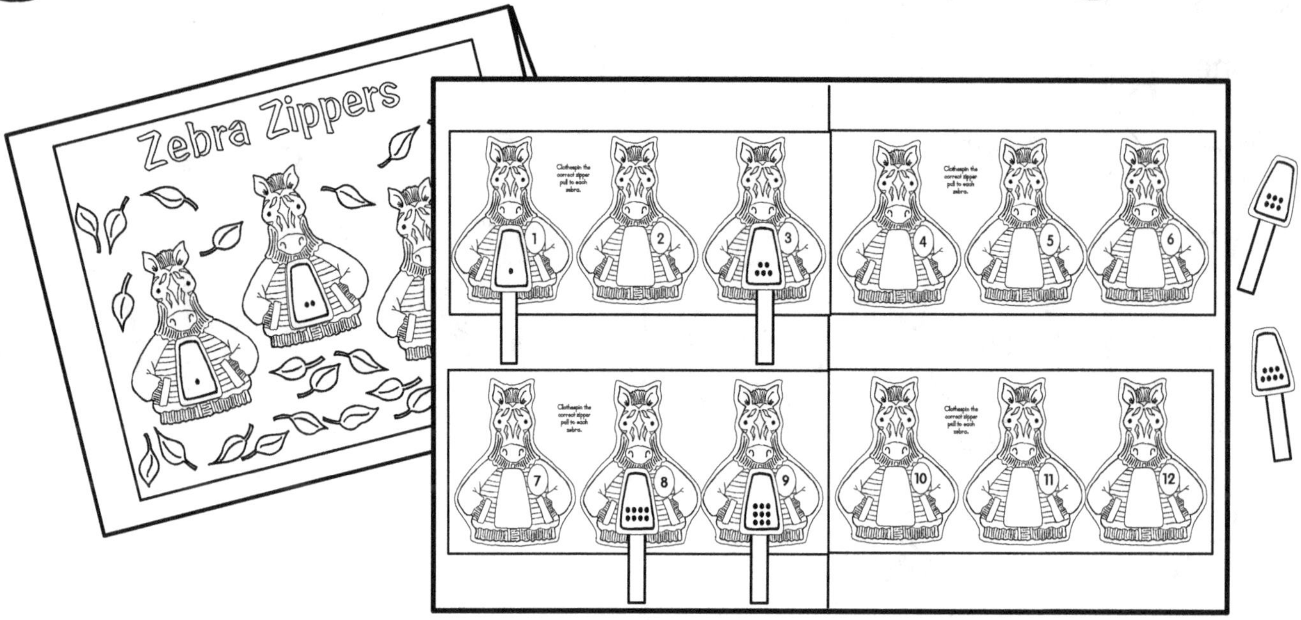

Clothespin Match-ups

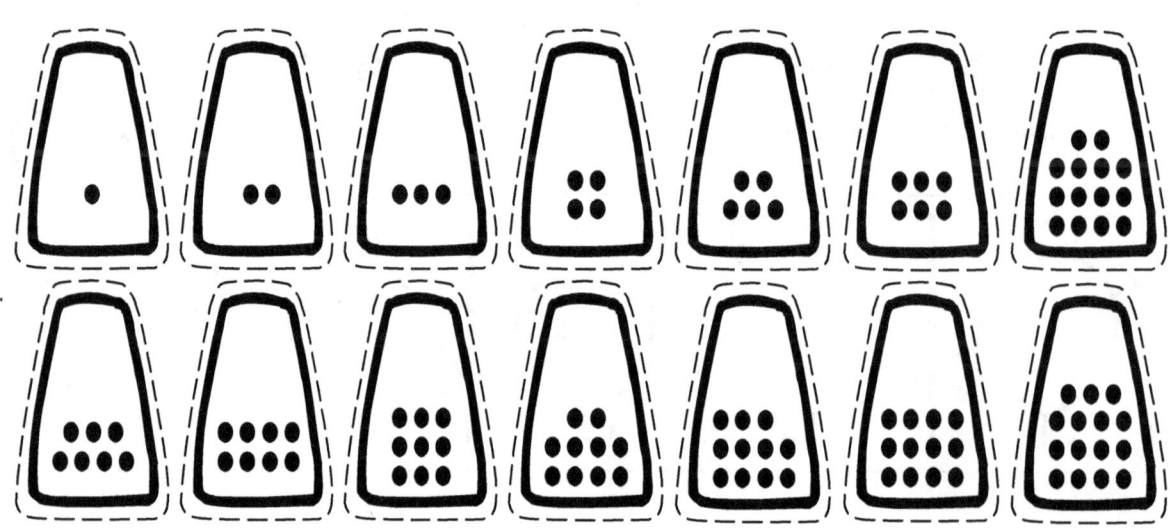

Assembly

Reproduce, cut out, and glue a set of oak tag handles (p. 192) to a standard-size file folder. Option: Color black stripes on the handles. Reproduce, color, and cut out the "Zebra Zippers" patterns. Glue the zipper pulls on clothespins. Glue the zebra cutouts in numerical order on zebra strips. Then glue the zebra strips to the inside of the folder. Glue the cover art to the front of the folder. Decorate the border around the zebra boards. Staple a resealable plastic bag to the back of the folder for clothespin storage. Option: Reproduce, cut out, and glue the Clothespin Match-ups directions to the back of the folder.

Zipper Pulls

Reproduce, cut apart, and glue a zipper pull on each clothespin. Include the two zipper pulls on the far right for advanced skills practice.

Zebra Zippers Cover
Sequencing Number Sets 1-12

Clothespin Match-ups

Zebra Zippers
Sequencing Number Sets 1-12

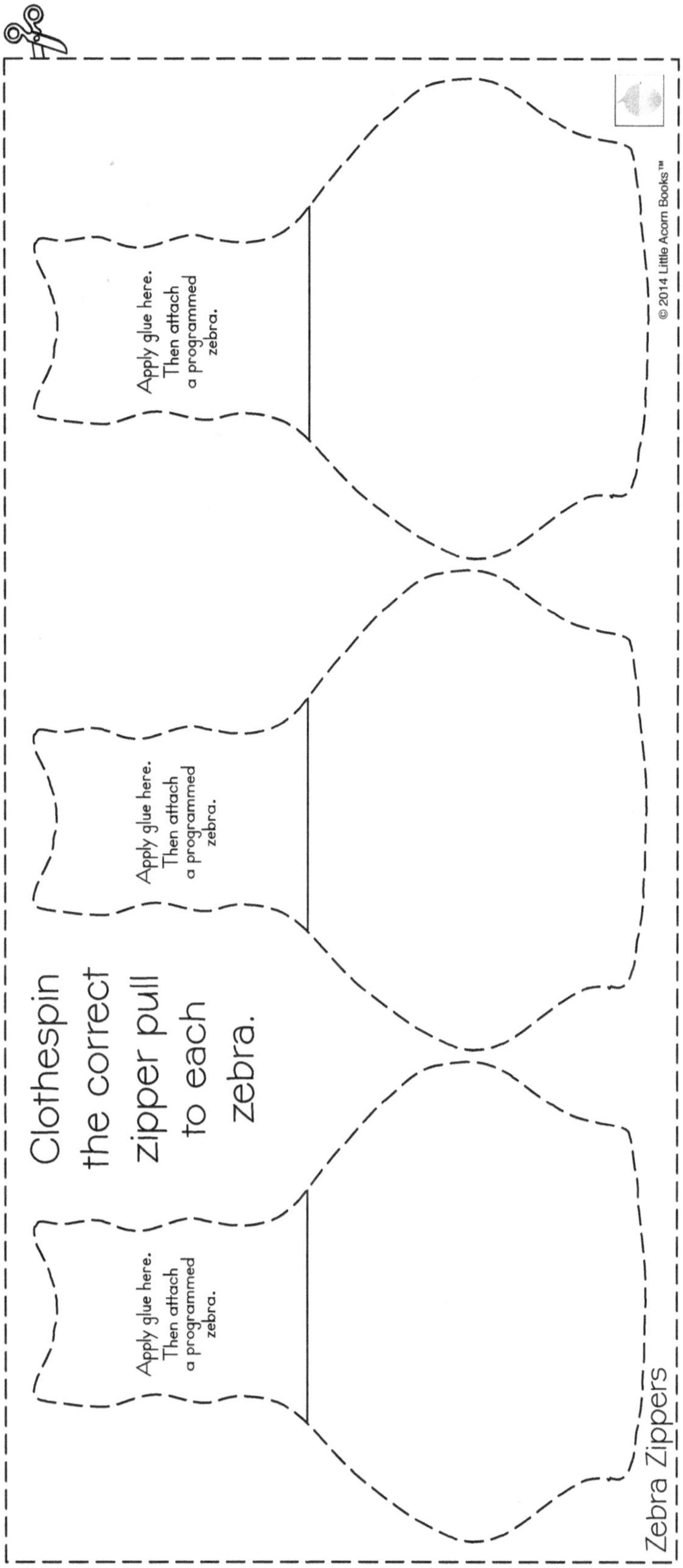

Clothespin Match-ups

Apply glue here. Then attach a programmed zebra.

Apply glue here. Then attach a programmed zebra.

Clothespin the correct zipper pull to each zebra.

Apply glue here. Then attach a programmed zebra.

Zebra Zippers

Reproduce, cut out, and glue four zebra strips to the inside of a folder.
Do not glue down bottom portion of zebras.

Zebra Zippers
Sequencing Number Sets 1-12

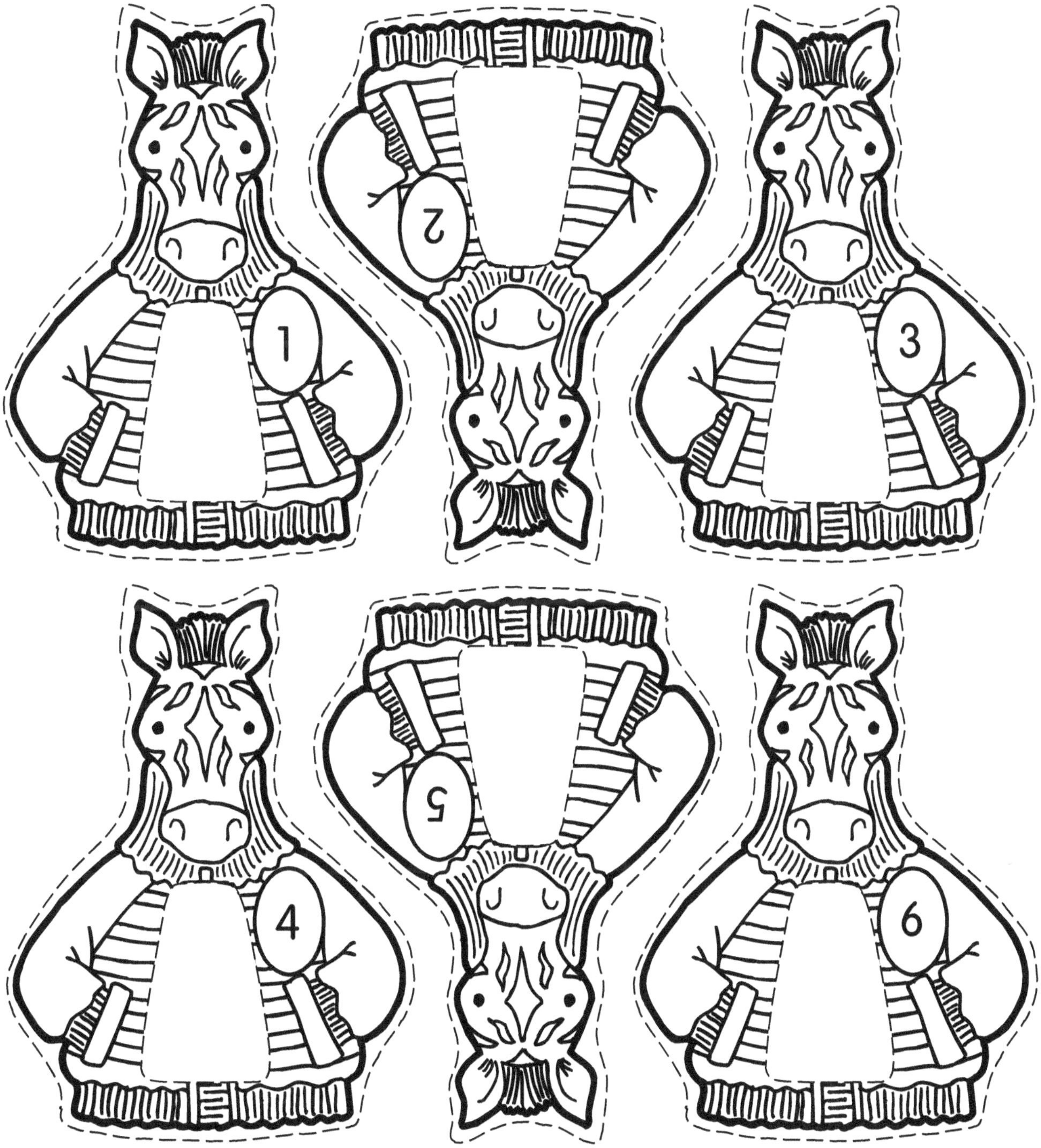

Clothespin Match-ups

Reproduce, color, and cut out. Arrange and glue the zebras in numerical order on zebra strips.

Zebra Zippers
Sequencing Number Sets 1-12

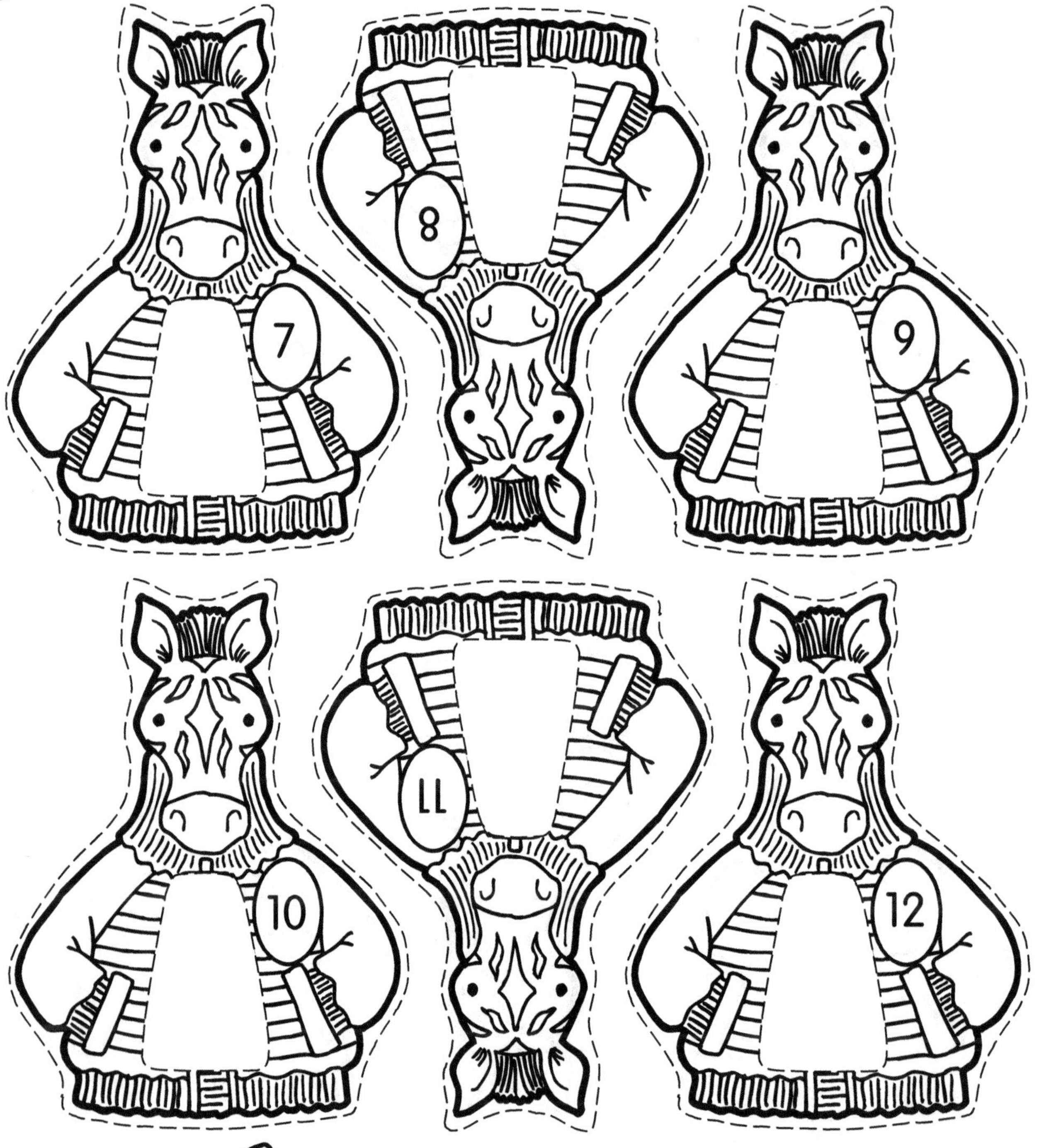

Clothespin Match-ups

Reproduce, color, and cut out. Arrange and glue the zebras in numerical order on zebra strips.

Great Goose Cover

Sequencing Numerals 1-12

Dot-to-dot

Assembly

Reproduce, cut out, and glue a set of oak tag handles (p. 192) to a standard-size file folder. Option: Decorate each handle with a copy of the geese on this page. Reproduce, color, and cut out the "Great Goose" patterns. Matching in the center, glue the goose to the inside of the folder. Glue the cover art to the front of the folder, then laminate. Laminate, then cut apart, the dot-to-dot pictures. Glue an envelope to the back of the folder to store dot-to-dot pictures, a wipe-off marker or crayon, and soft cloth. Children take out the pictures one at a time to place on the nest. Option: Reproduce, cut out, and glue the Dot-to-Dot Playing Directions to the back of the folder.

Great Goose Cover

Sequencing Numerals 1-12

Dot-to-dot

Great Goose
Sequencing Numerals 1-12

Dot-to-dot

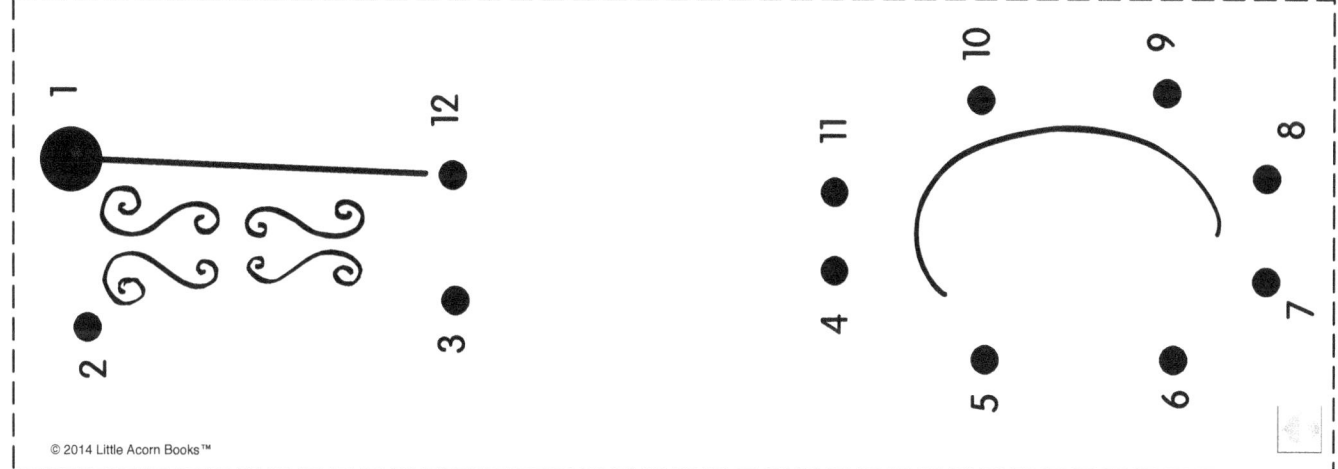

Spoon

Boat

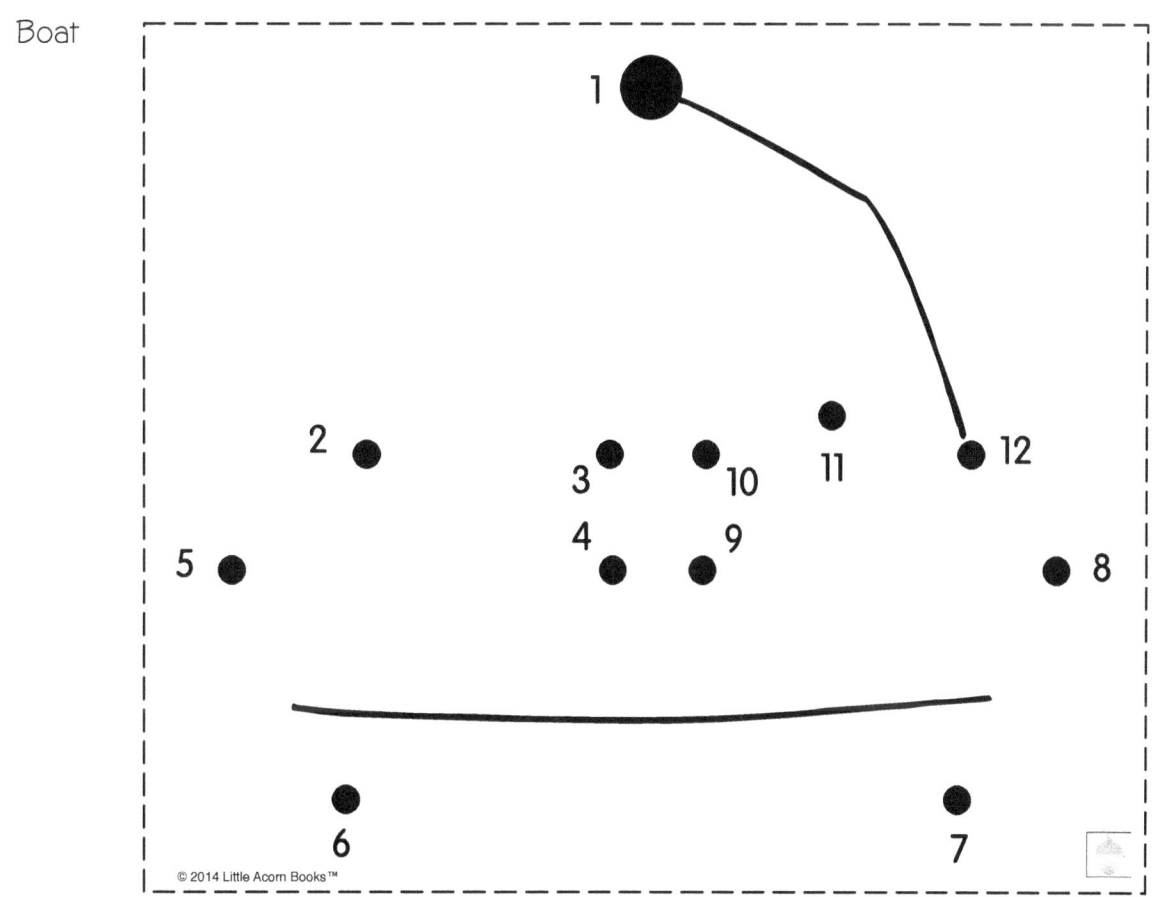

Great Goose
Sequencing Numerals 1-12

Dot-to-dot

Reproduce, color, cut out, and glue to the inside of a folder.

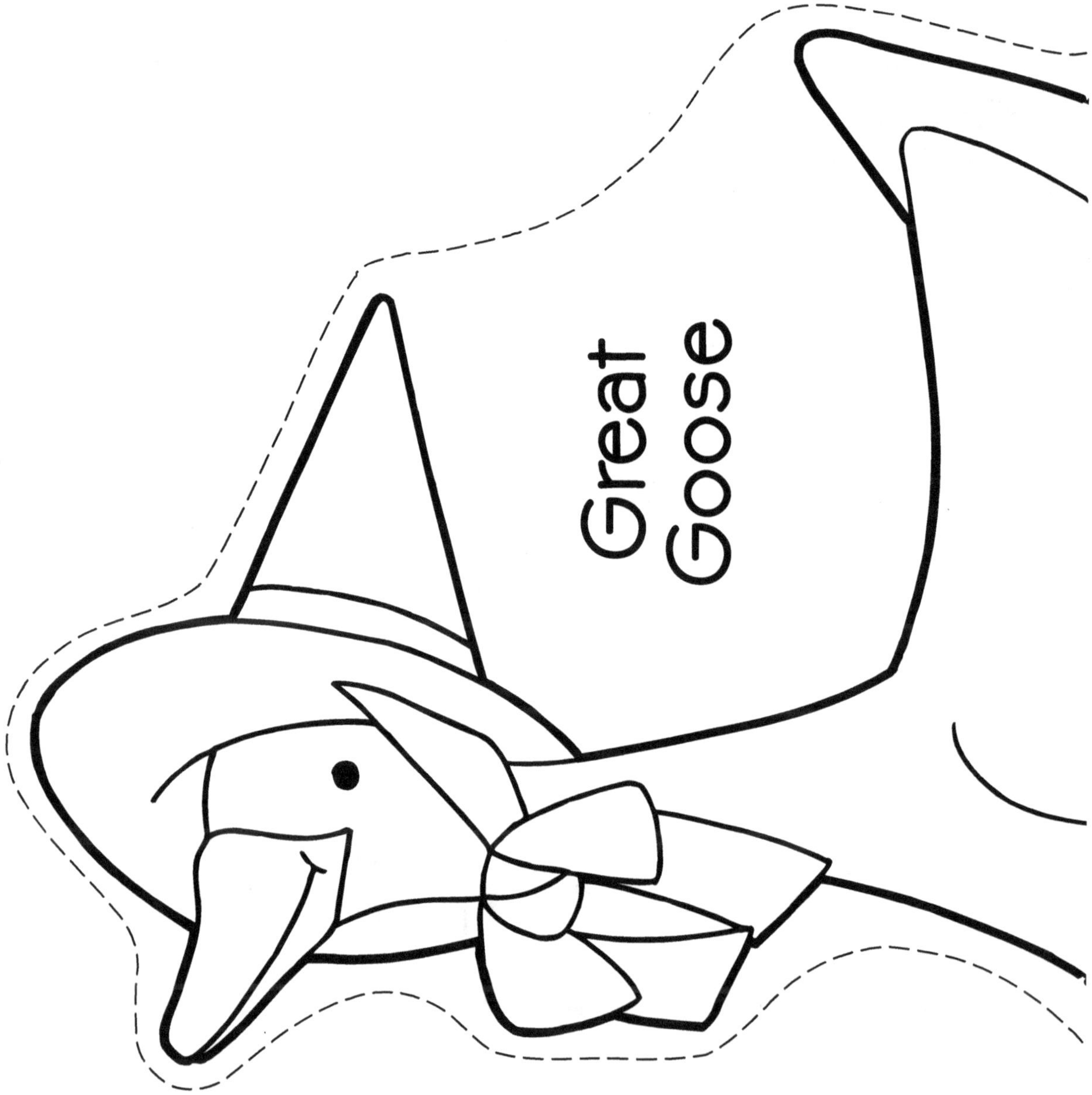

Great Goose

Start at the large dot on each picture.
Follow the numbers to connect the dots.

Great Goose
Sequencing Numerals 1-12

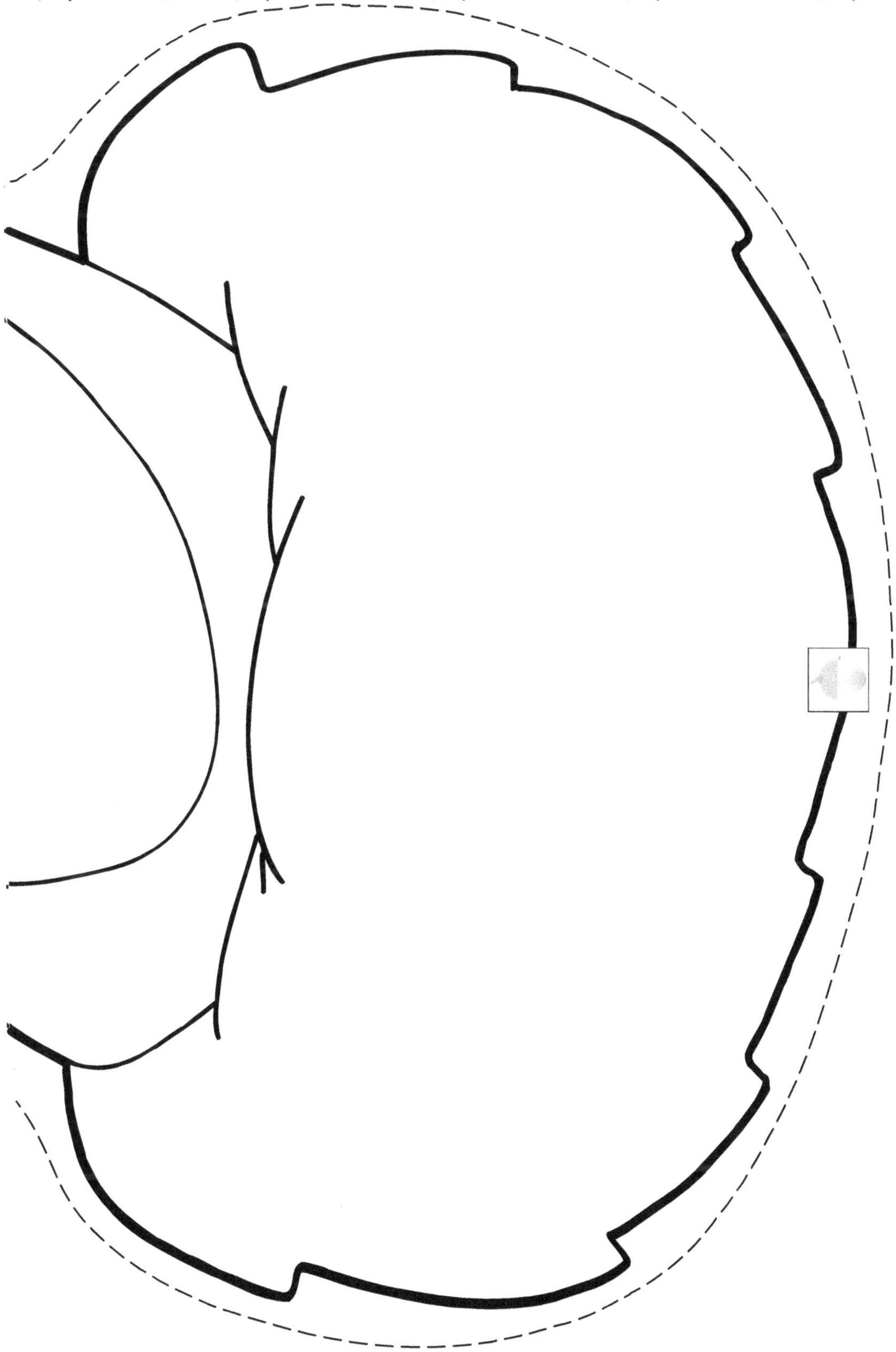

Dot-to-dot

Reproduce, color, cut out, and glue to the inside of a folder.

Great Goose
Sequencing Numerals 1-12

Dot-to-dot

Mask

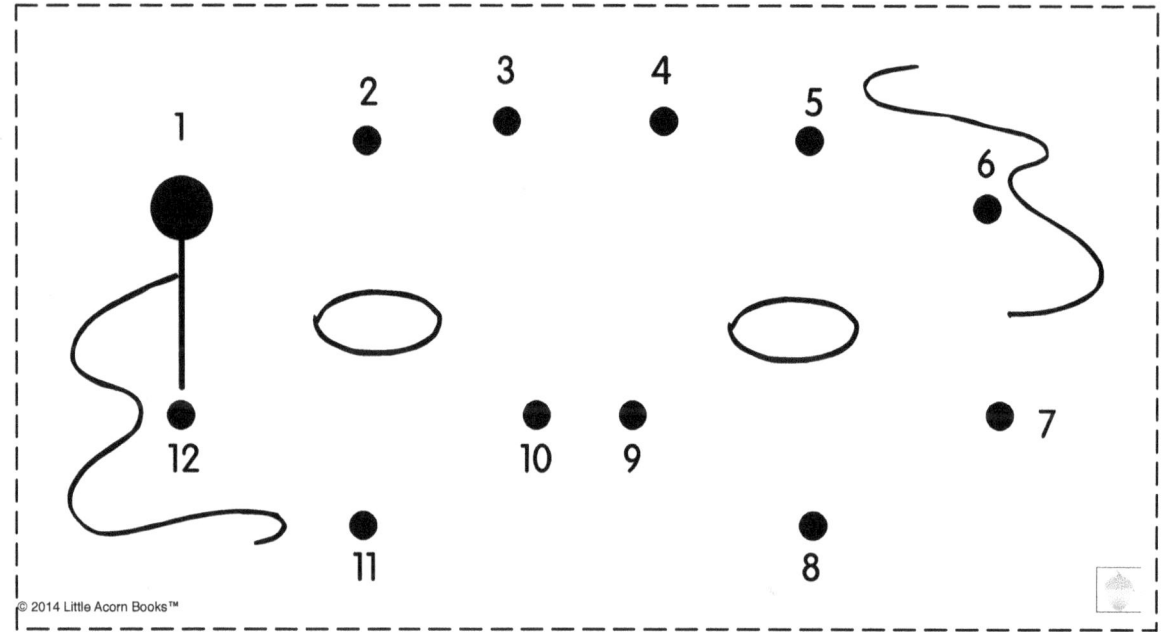

Fish

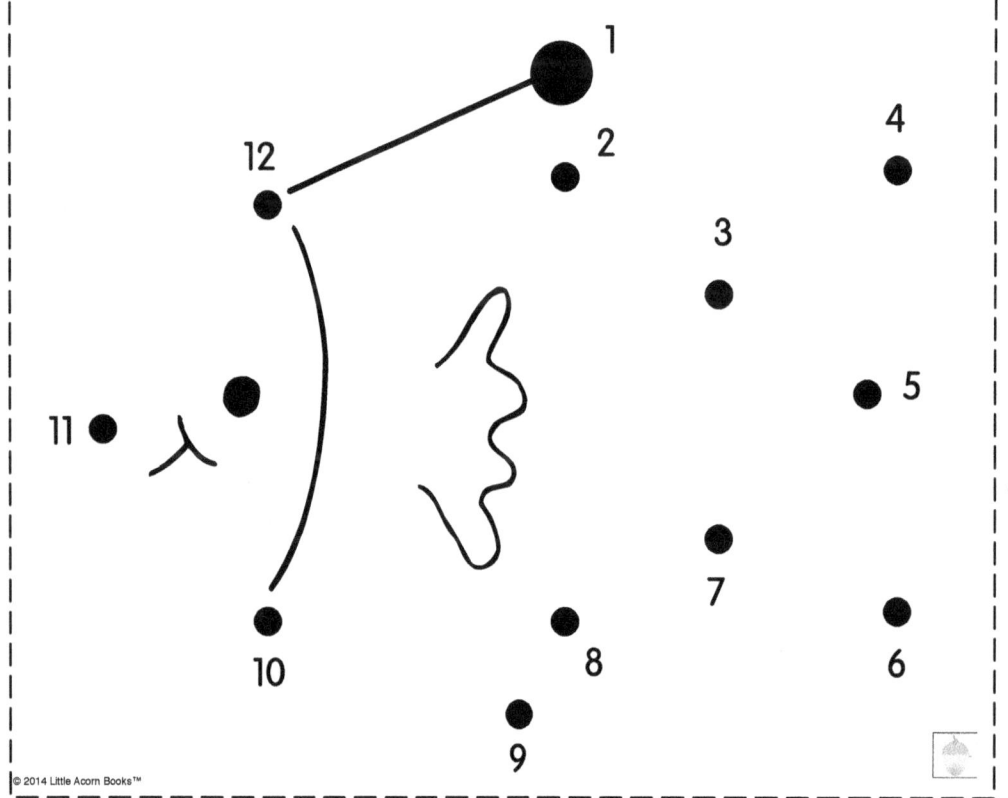

Peek-a-boo Penguins

Pocket Game

Sequencing Numerals and Number Sets 1-12

Assembly

Reproduce, cut out, and glue a set of oak tag handles (p. 192) to a standard-size file folder. Option: Decorate each handle with a copy of the penguins on this page. Reproduce, color, and cut out the "Peek-a-boo Penguins" patterns. Glue two placement templates to the inside of the folder. Apply glue only along the bottom and side edges, then attach igloos to each template. Glue the cover art to the front of the folder. Decorate the border around the game board. Glue an envelope to the back of the folder for penguin storage. Option: Reproduce, cut out, and glue the Pocket Game Playing Directions to the back of the folder. Note: Make additional "Peek-a-boo Penguins" pocket folders with different combinations of baskets.

Directions

Reproduce, cut out, and glue the direction strip to the game board.

Place each penguin in a matching igloo.

Peek-a-boo Penguins Cover
Sequencing Numerals and Number Sets 1-12

Pocket Game

Peek-a-boo Penguins

Peek-a-boo Penguins
Sequencing Numerals and Number Sets 1-12

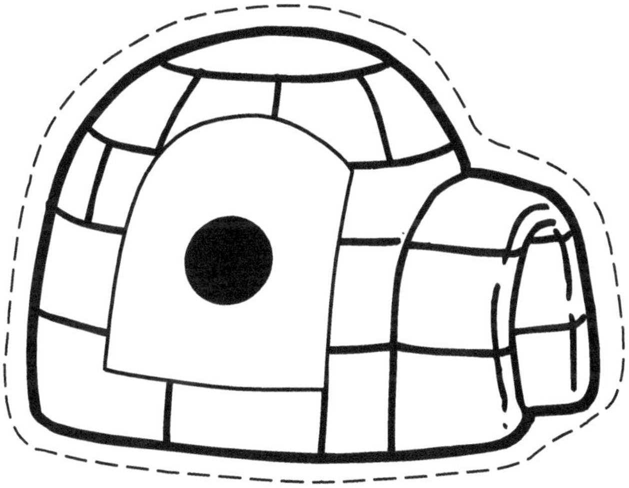

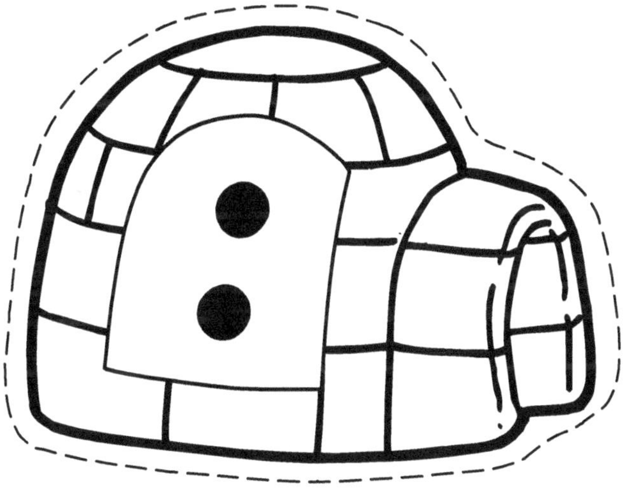

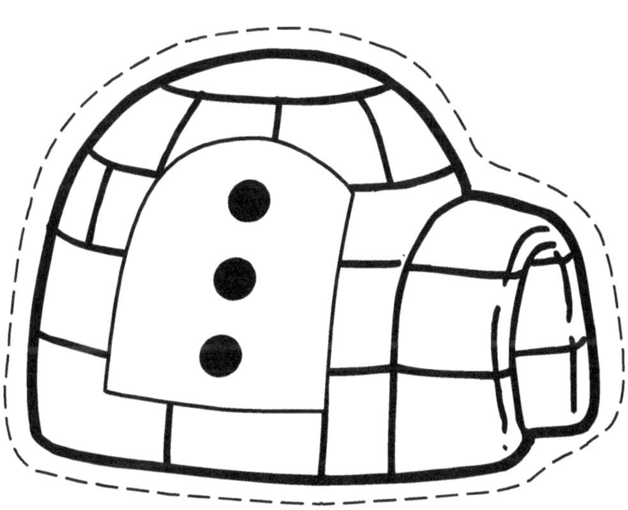

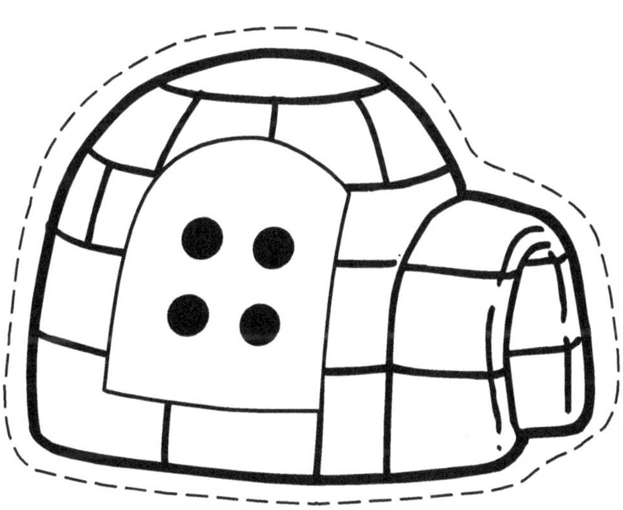

Pocket Game

Reproduce, color, and cut out the igloos. Apply glue only along the bottom and side edges of each igloo. Attach each igloo to the inside of a folder.

Peek-a-boo Penguins
Sequencing Numerals and Number Sets 1-12

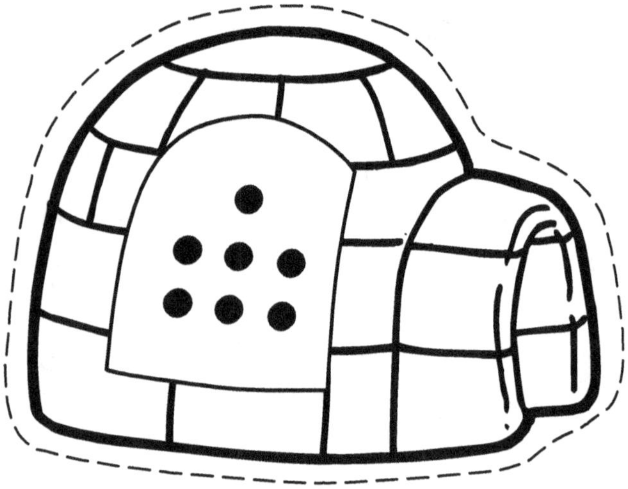

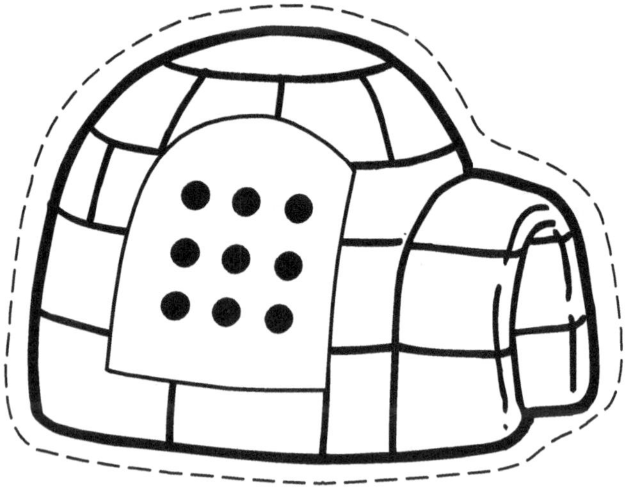

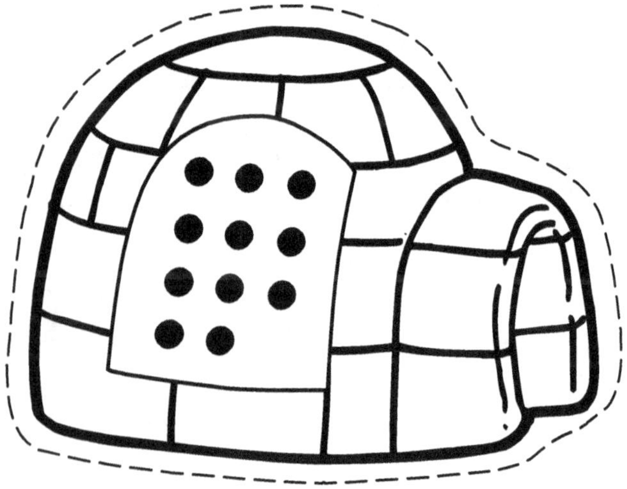

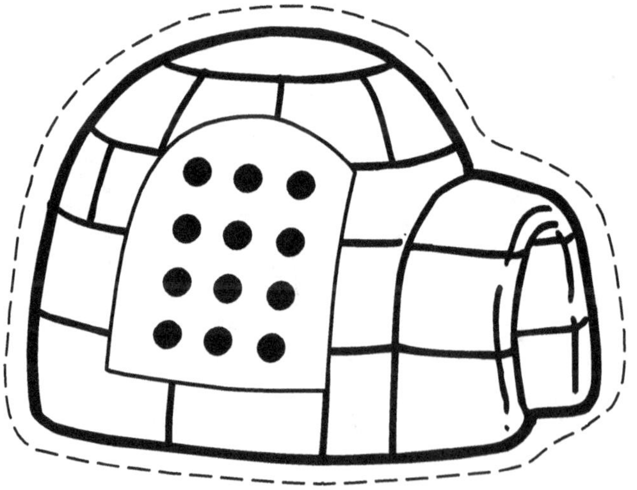

Pocket Game

Reproduce, color, and cut out the igloos. Apply glue only along the bottom and side edges of each igloo. Attach each igloo to the inside of a folder.

Peek-a-boo Penguins
Sequencing Numerals and Number Sets 1-12

Pocket Game

1	2	3	4
5	6	7	8
9	10	11	12

LAB20145P • 35 FOLDERGAMES FOR NUMBERS • 978-1-937257-53-8 • © 2014 Little Acorn Books™   155

Peek-a-boo Roos

Matching Number Sets and Number Words 1-12

Pocket Game

Assembly

Reproduce, cut out, and glue a set of oak tag handles (p. 192) to a standard-size file folder. Option: Decorate each handle with a copy of the kangaroos above. Reproduce, color, and cut out the "Peek-a-boo Roos" patterns. Glue the kangaroo boards to the inside of the folder. Apply glue only along the curved edge of each pouch. Attach the pouches to kangaroo boards. Glue the cover art to the front of the folder. Decorate the border around the game board. Glue an envelope to the back of the folder for joey storage. Option: Reproduce, cut out, and glue the Pocket Game Playing Directions to the back of the folder. Note: Make additional "Peek-a-boo Roos" pocket folders for additional matching skills practice. Reproduce and program non-matching roos for advanced skills practice.

Peek-a-boo Roos Cover
Matching Number Sets and Number Words 1-12

Pocket Game

Peek-a-boo Roos

Pocket Game

Matching Number Sets and Number Words 1-12

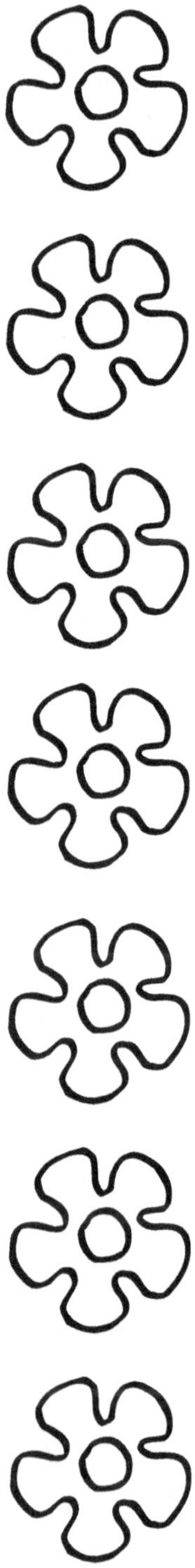

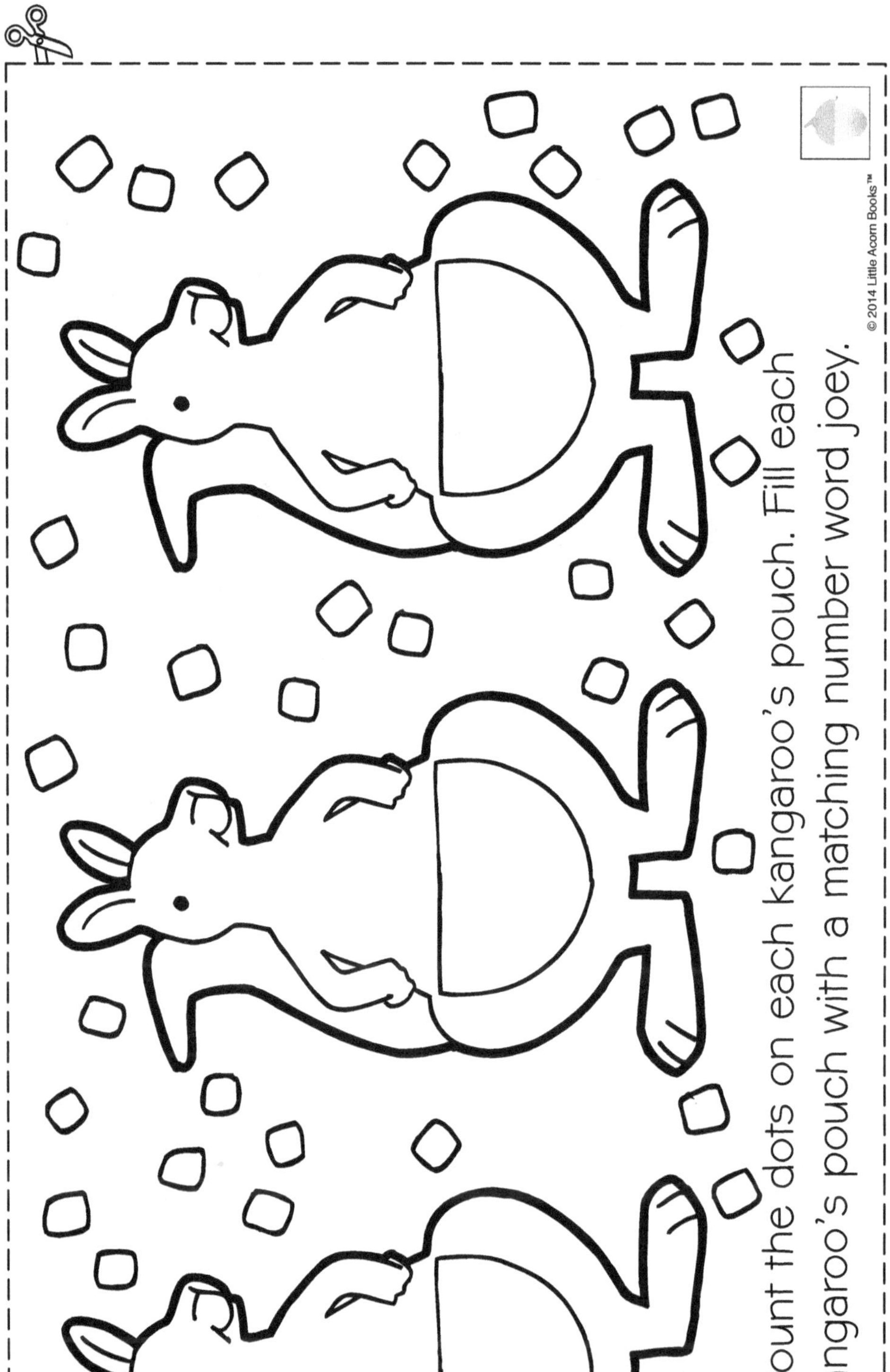
Count the dots on each kangaroo's pouch. Fill each kangaroo's pouch with a matching number word joey.
Peek-a-boo Roos
© 2014 Little Acorn Books™

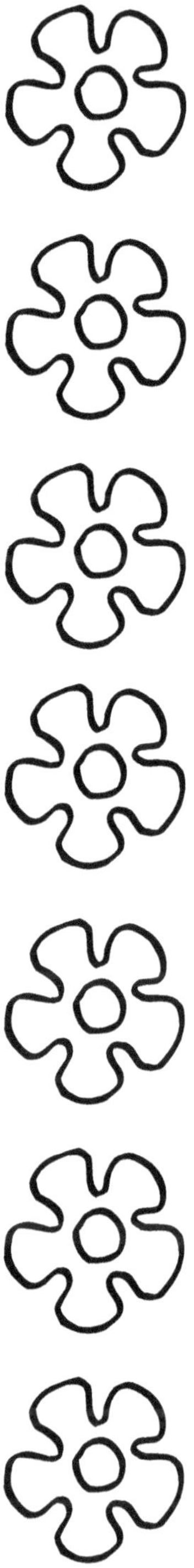

Peek-a-boo Roos

Pocket Game

Matching Number Sets and Number Words 1-12

Reproduce, color, and cut out.
Apply glue only along the curved edge of each pouch.

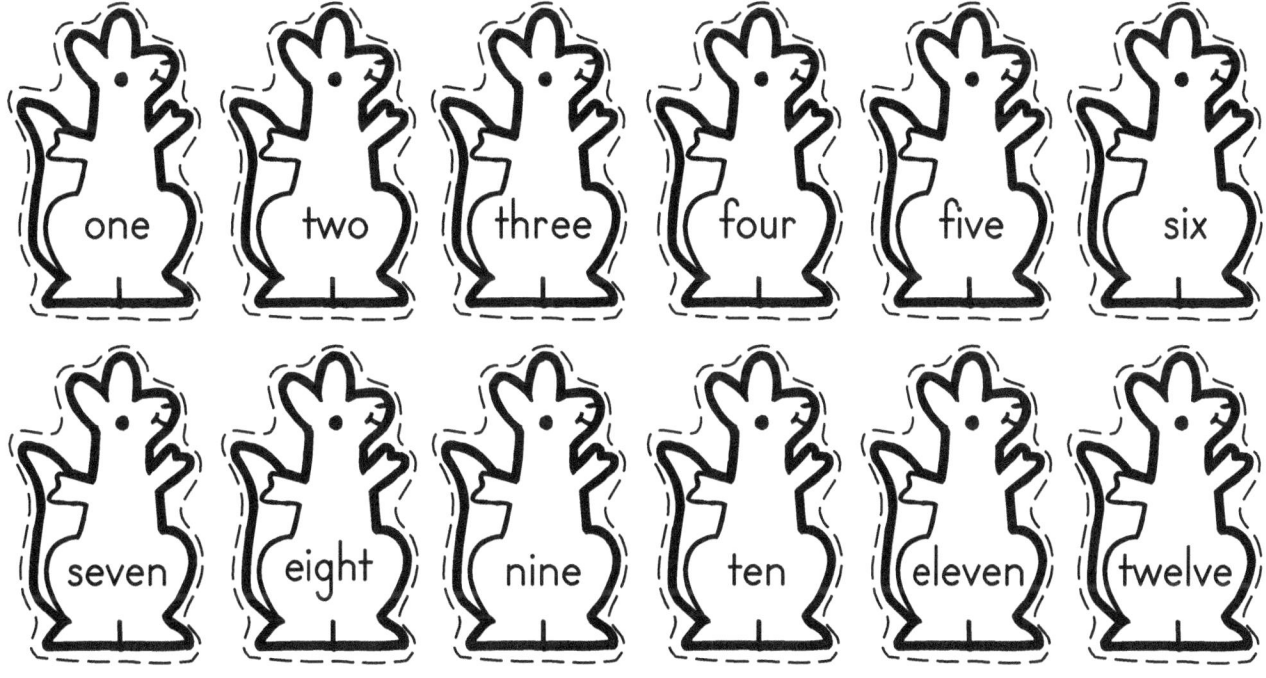

Attach the pouches to kangaroos on Peek-a-boo Roos pocket boards.

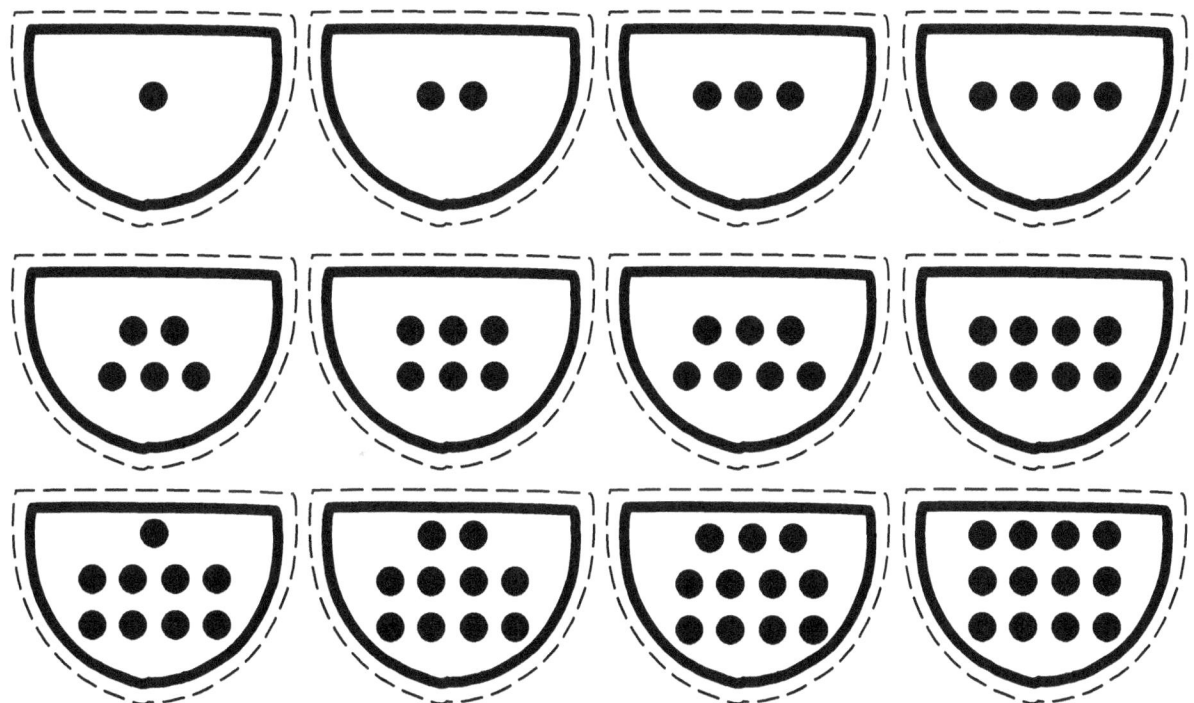

Dancing Dragons
Matching Numerals and Number Words 1-12

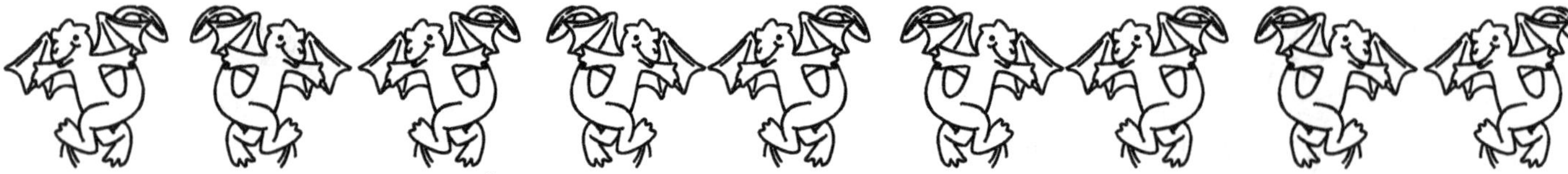

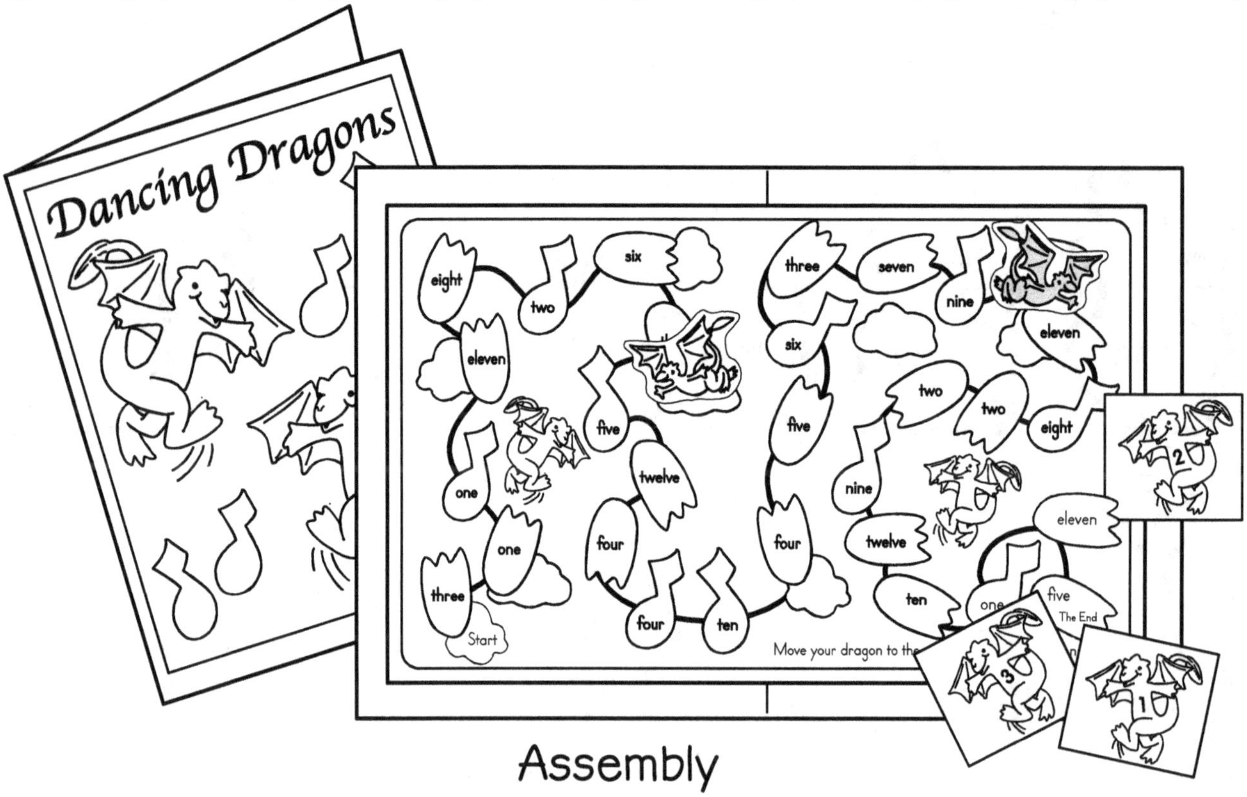

Trail Game

Assembly

Reproduce, cut out, and glue a set of oak tag handles (p. 192) to a standard-size file folder. Option: Decorate each handle with a copy of the dragons on this page. Reproduce, color, and cut out the "Dancing Dragons" patterns. Matching in the center, glue the game board patterns to the inside of the folder. Glue the cover art to the front of the folder. Decorate the border around the game board, then laminate. Reproduce, color, laminate, then cut out the pawns and two sets of game cards. Glue an envelope to the back of the folder for pawn and game card storage. Option: Reproduce, cut out, and glue the Trail Game Playing Directions to the back of the folder.

Pawns
Reproduce, color, and cut out.

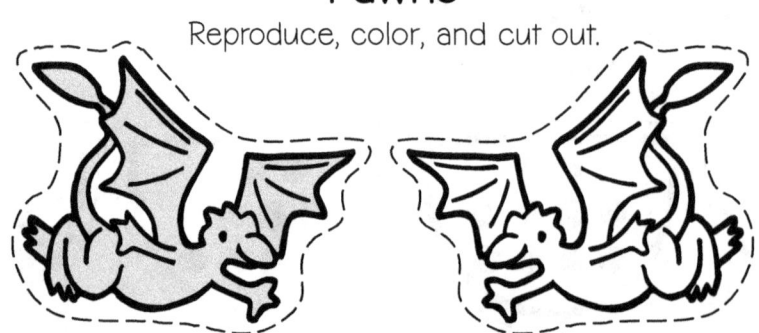

Dancing Dragons Cover

Matching Numerals and Number Words 1-12

Trail Game

Dancing Dragons

Dancing Dragons
Matching Numerals and Number Words 1-12

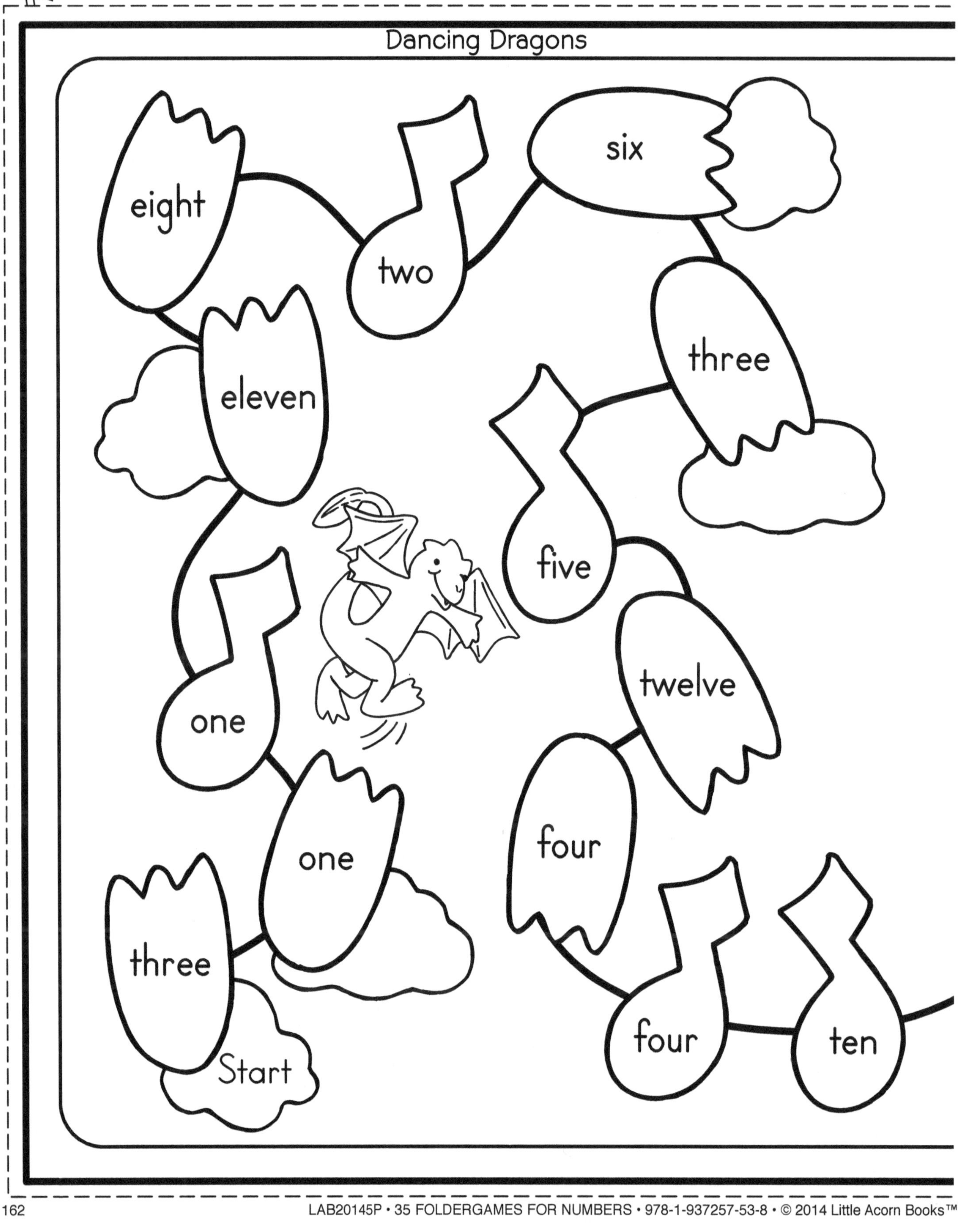

Trail Game

Dancing Dragons

eight • two • six • three • eleven • five • one • twelve • one • four • three • four • ten • Start

Dancing Dragons

Matching Numerals and Number Words 1-12

Trail Game

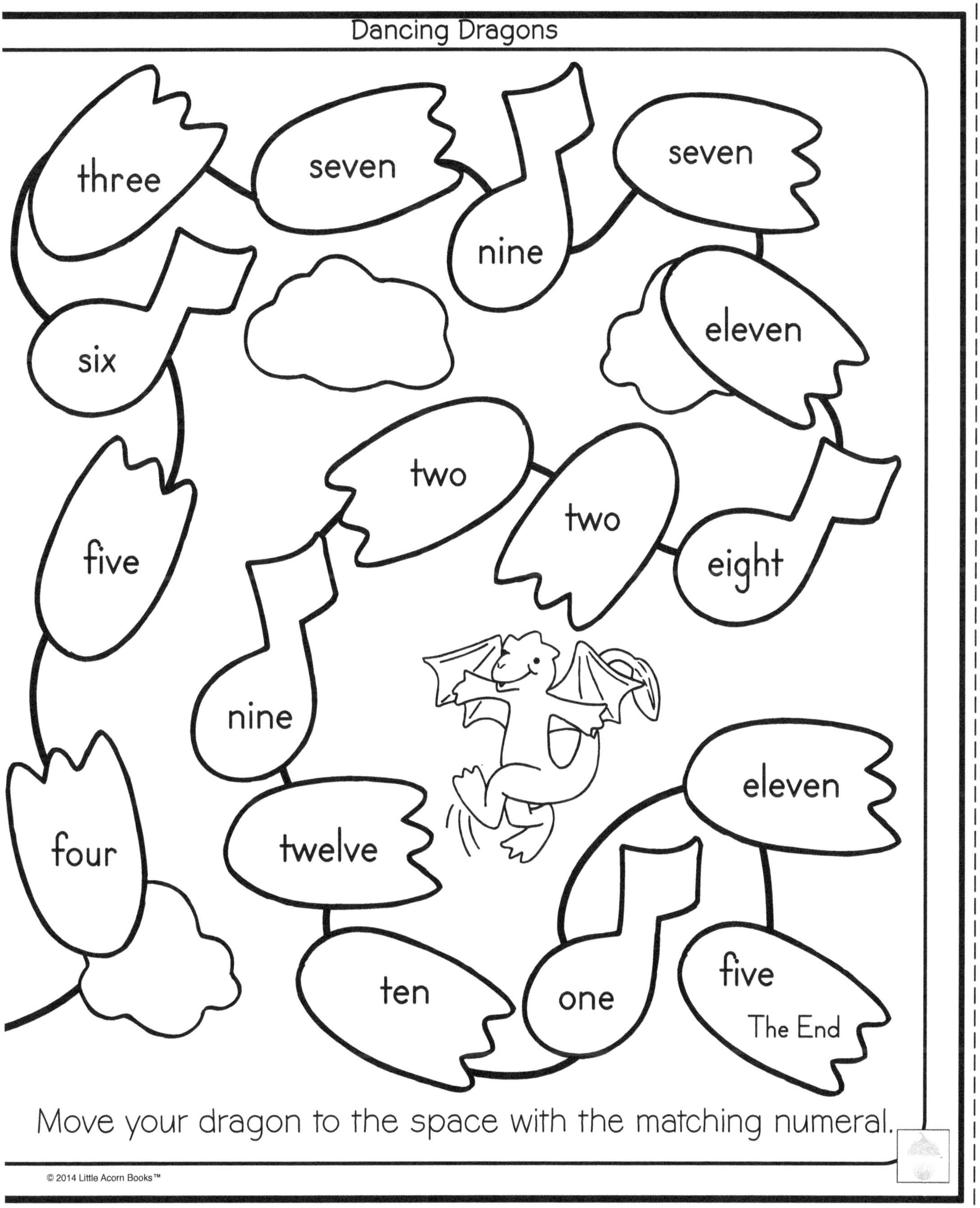
Dancing Dragons
three
seven
nine
seven
six
eleven
two
five
two
eight
nine
eleven
four
twelve
ten
one
five
The End
Move your dragon to the space with the matching numeral.
© 2014 Little Acorn Books™

Trail Game

Dancing Dragons

Matching Numerals and Number Words 1-12

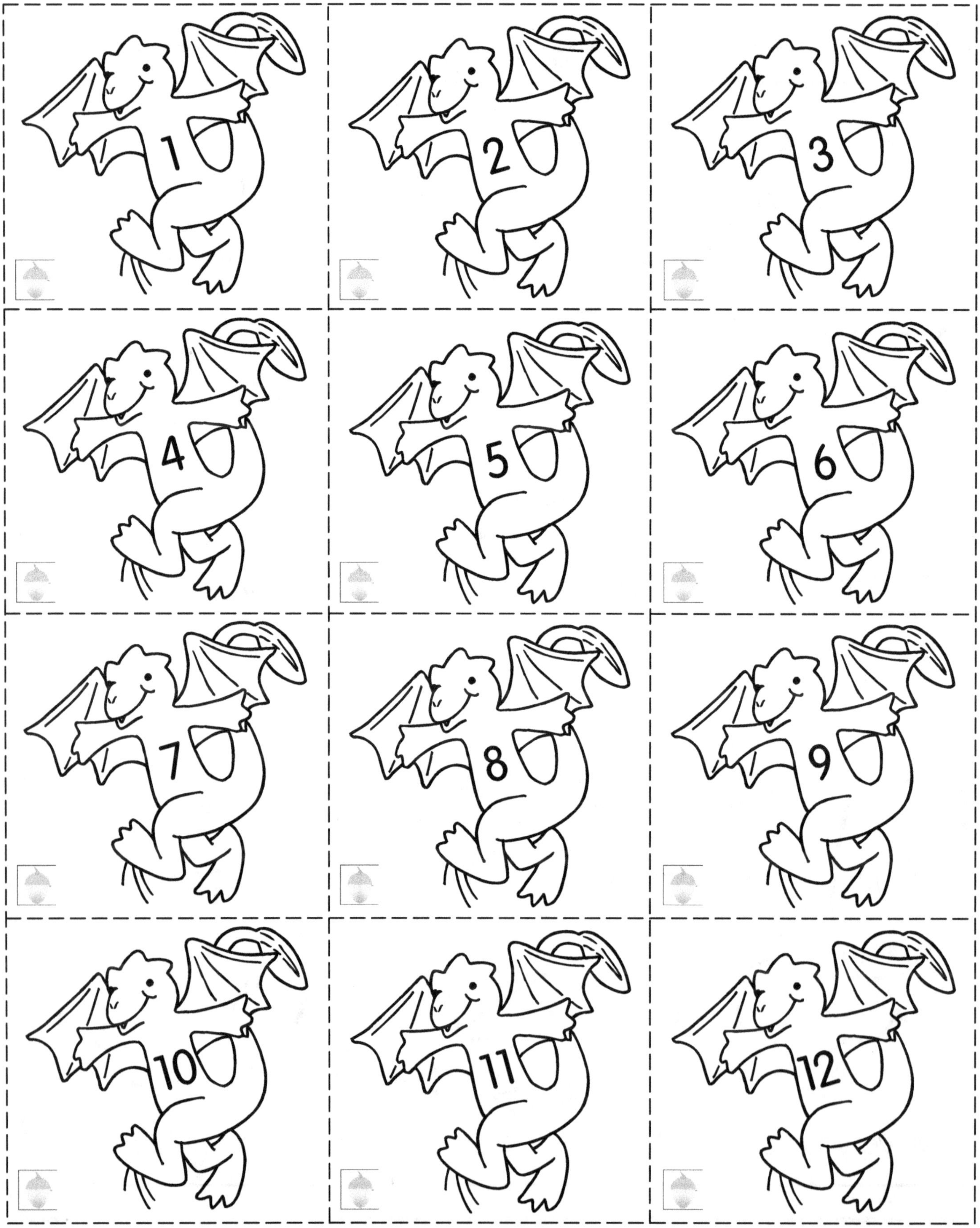

LAB20145P • 35 FOLDERGAMES FOR NUMBERS • 978-1-937257-53-8 •

Peek-a-boo Fish
Telling Time

Pocket Game

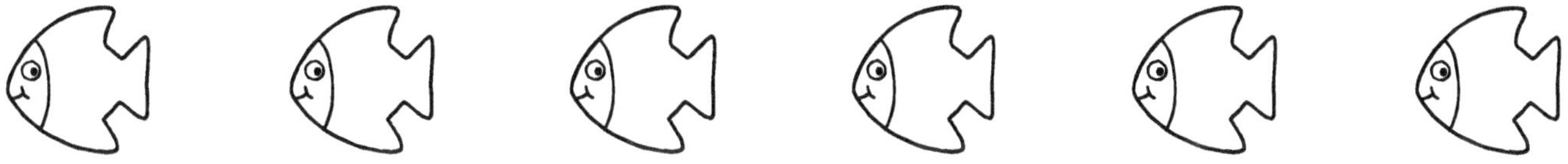

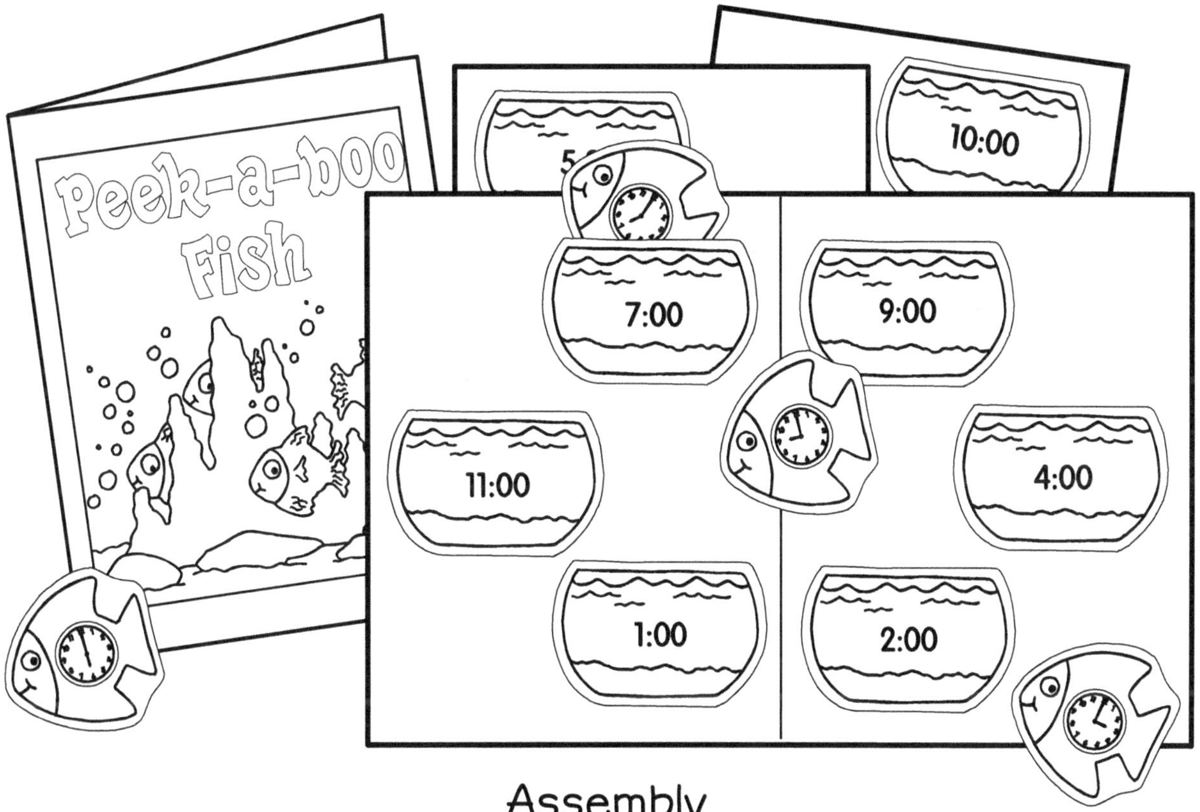

Assembly

Reproduce, cut out, and glue a set of oak tag handles (p. 192) to a standard-size file folder. Option: Decorate each handle with a copy of the fish on this page. Reproduce, color, and cut out the "Peek-a-boo Fish" patterns. Apply glue only along the bottom and side edges, then attach three fish bowls to each inside panel of the folder. Glue the rest of the bowls to two separate sheets of oak tag. Glue the cover art to the front of the folder. Decorate the border around the game board. Glue an envelope to the back of the folder for fish storage. Option: Reproduce, cut out, and glue the Pocket Game Playing Directions to the back of the folder.

Directions

Reproduce, cut out, and glue the direction strip to the game board.

Place each fish in a matching bowl.

Peek-a-boo Fish Cover
Telling Time

Pocket Game

Peek-a-boo Fish
Telling Time

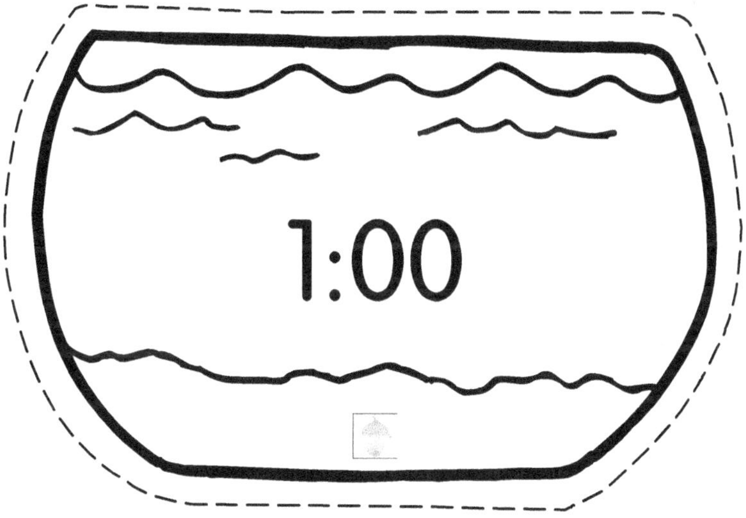

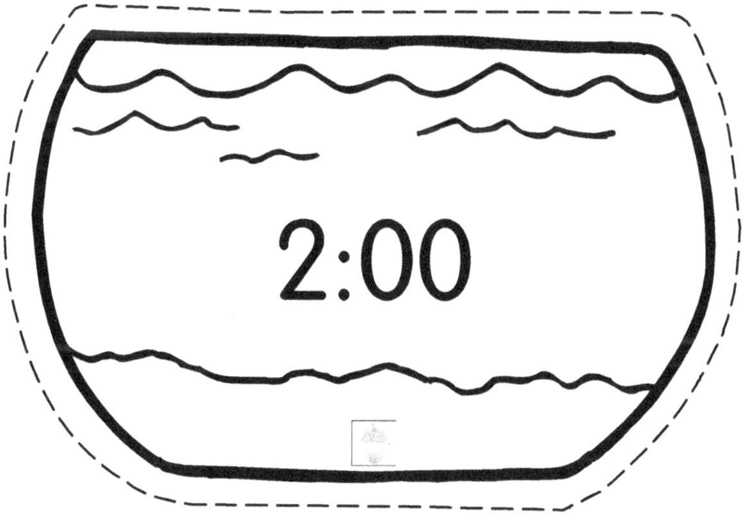

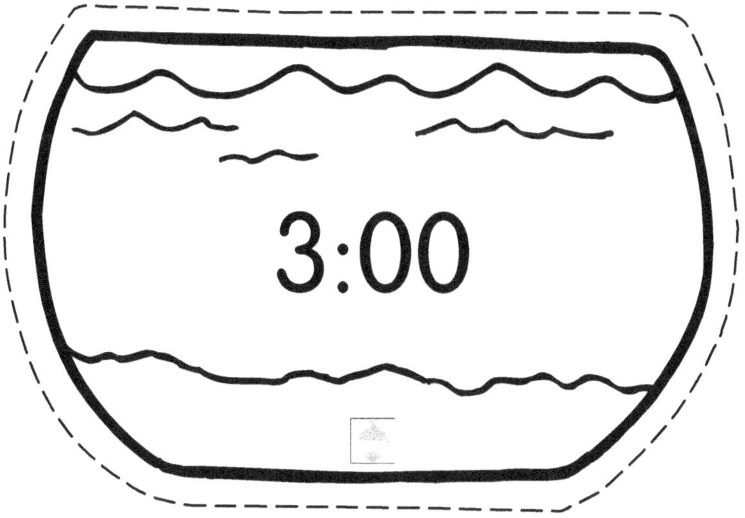

Pocket Game

Reproduce, color, and cut out the fish and bowls. Apply glue only along the bottom and side edges of each bowl. Attach each bowl to the inside of a folder.

Peek-a-boo Fish
Telling Time

Pocket Game

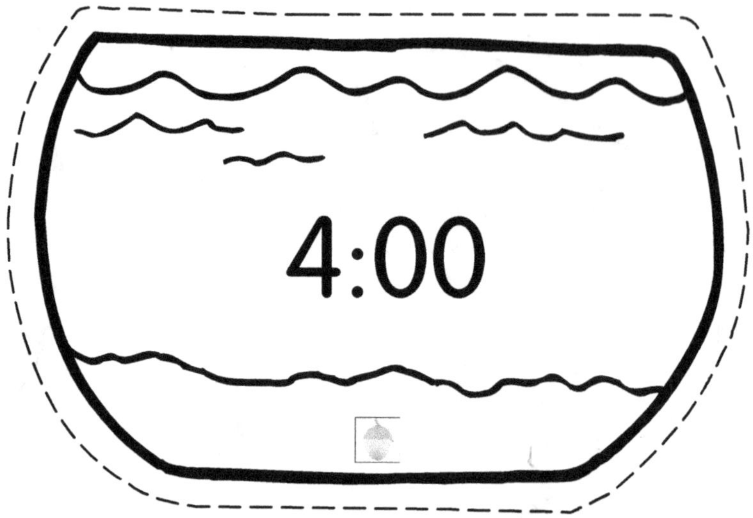

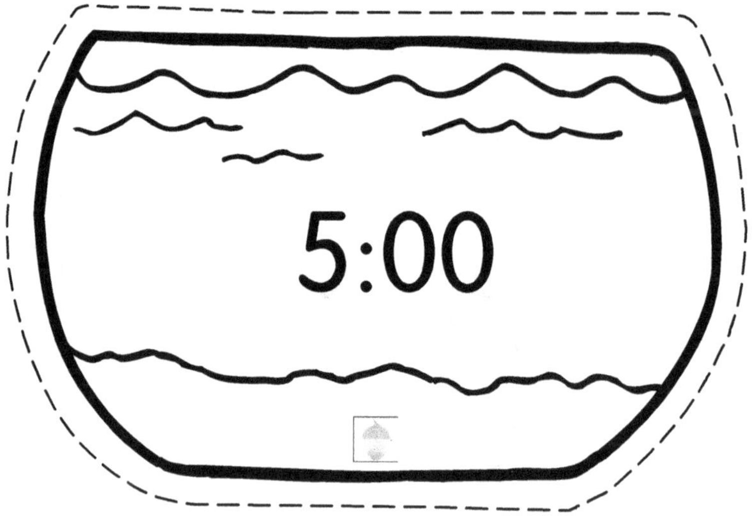

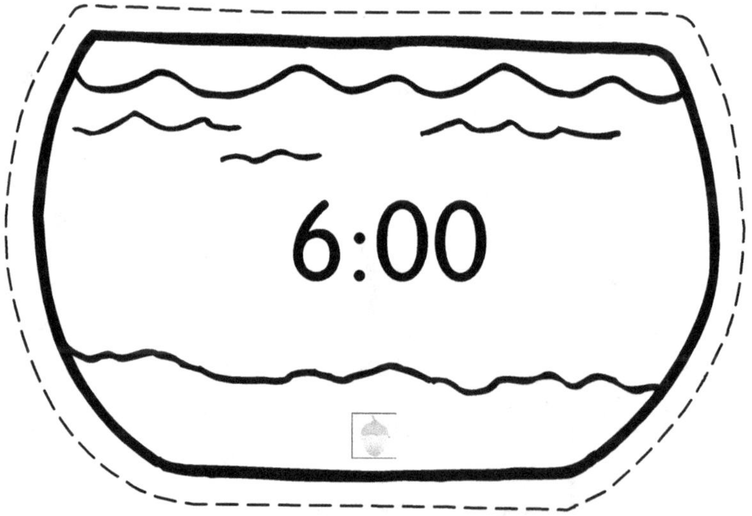

Reproduce, color, and cut out the fish and bowls. Apply glue only along the bottom and side edges of each bowl. Attach each bowl to the inside of a folder.

Peek-a-boo Fish
Telling Time

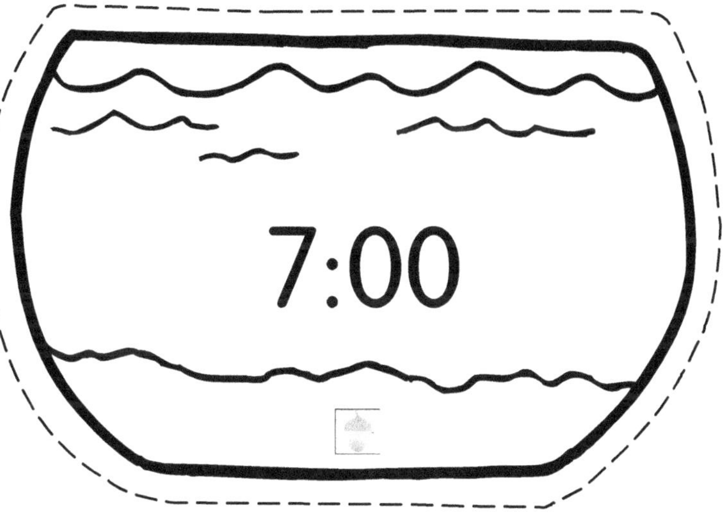

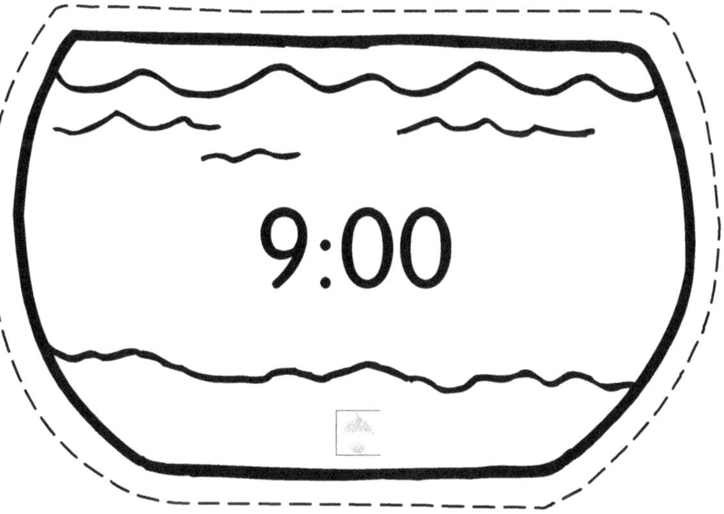

Pocket Game

7:00

8:00

9:00

Reproduce, color, and cut out the fish and bowls. Apply glue only along the bottom and side edges of each bowl. Attach each bowl to the inside of a folder.

Peek-a-boo Fish
Telling Time

Pocket Game

Reproduce, color, and cut out the fish and bowls. Apply glue only along the bottom and side edges of each bowl. Attach each bowl to the inside of a folder.

Sneaky Snakes
Telling Time

Trail Game

Assembly

Reproduce, cut out, and glue a set of oak tag handles (p. 192) to a standard-size file folder. Option: Decorate each handle with a copy of the snakes on this page. Reproduce, color, and cut out the "Sneaky Snakes" patterns. Matching in the center, glue the game board patterns to the inside of the folder. Glue the cover art to the front of the folder. Decorate the border around the game board, then laminate. Reproduce, color, laminate, then cut out the pawns and two sets of game cards. Glue an envelope to the back of the folder for pawn and game card storage. Option: Reproduce, cut out, and glue the Trail Game Playing Directions to the back of the folder.

Pawns
Reproduce, color, and cut out.

Sneaky Snakes
Telling Time

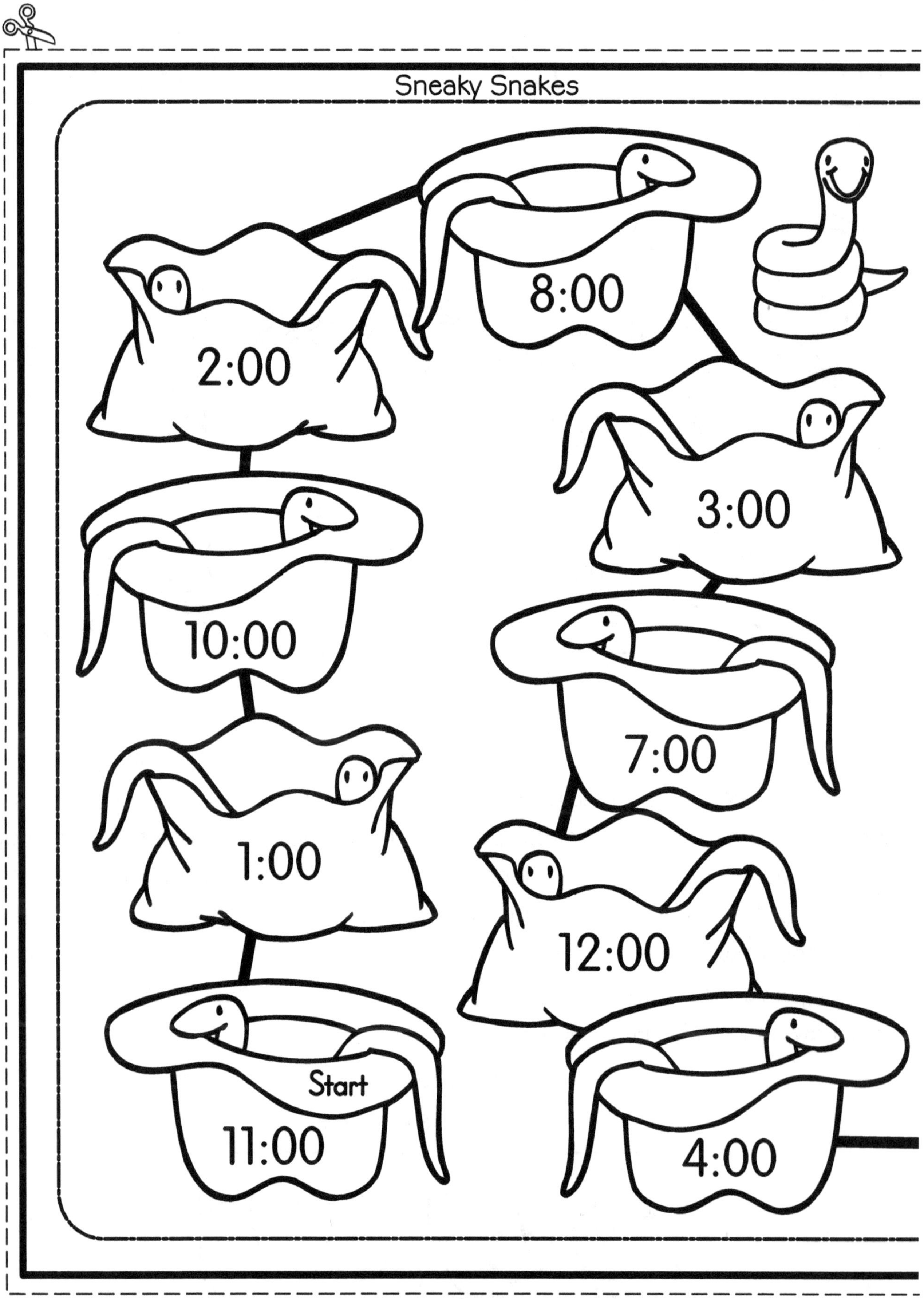

Trail Game

Sneaky Snakes

- 8:00
- 2:00
- 3:00
- 10:00
- 7:00
- 1:00
- 12:00
- Start 11:00
- 4:00

Sneaky Snakes
Telling Time

Trail Game

Sneaky Snakes

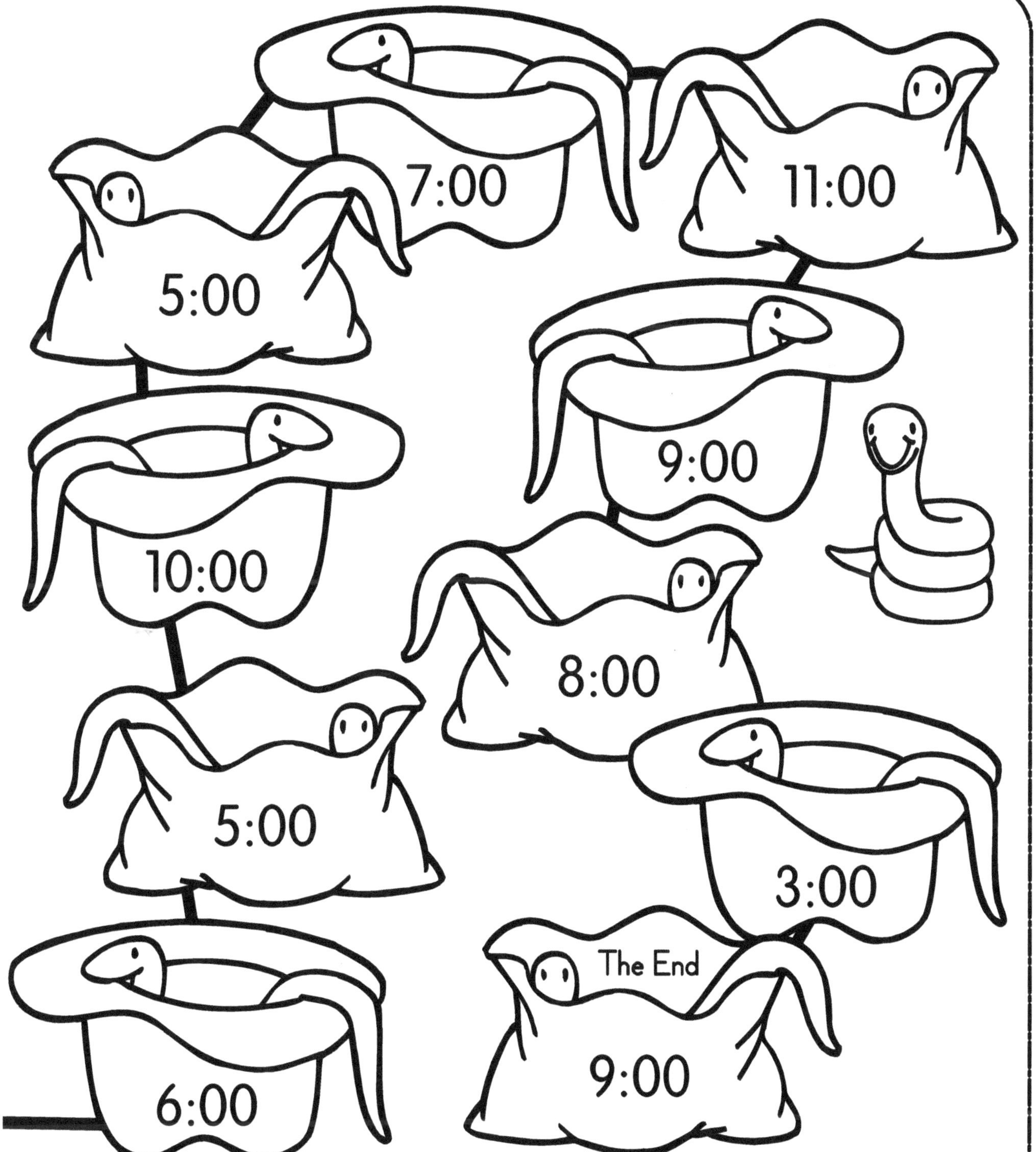

7:00
11:00
5:00
9:00
10:00
8:00
5:00
3:00
6:00
The End
9:00

Move your snake to the space with the matching time.

Sneaky Snakes
Telling Time

Trail Game

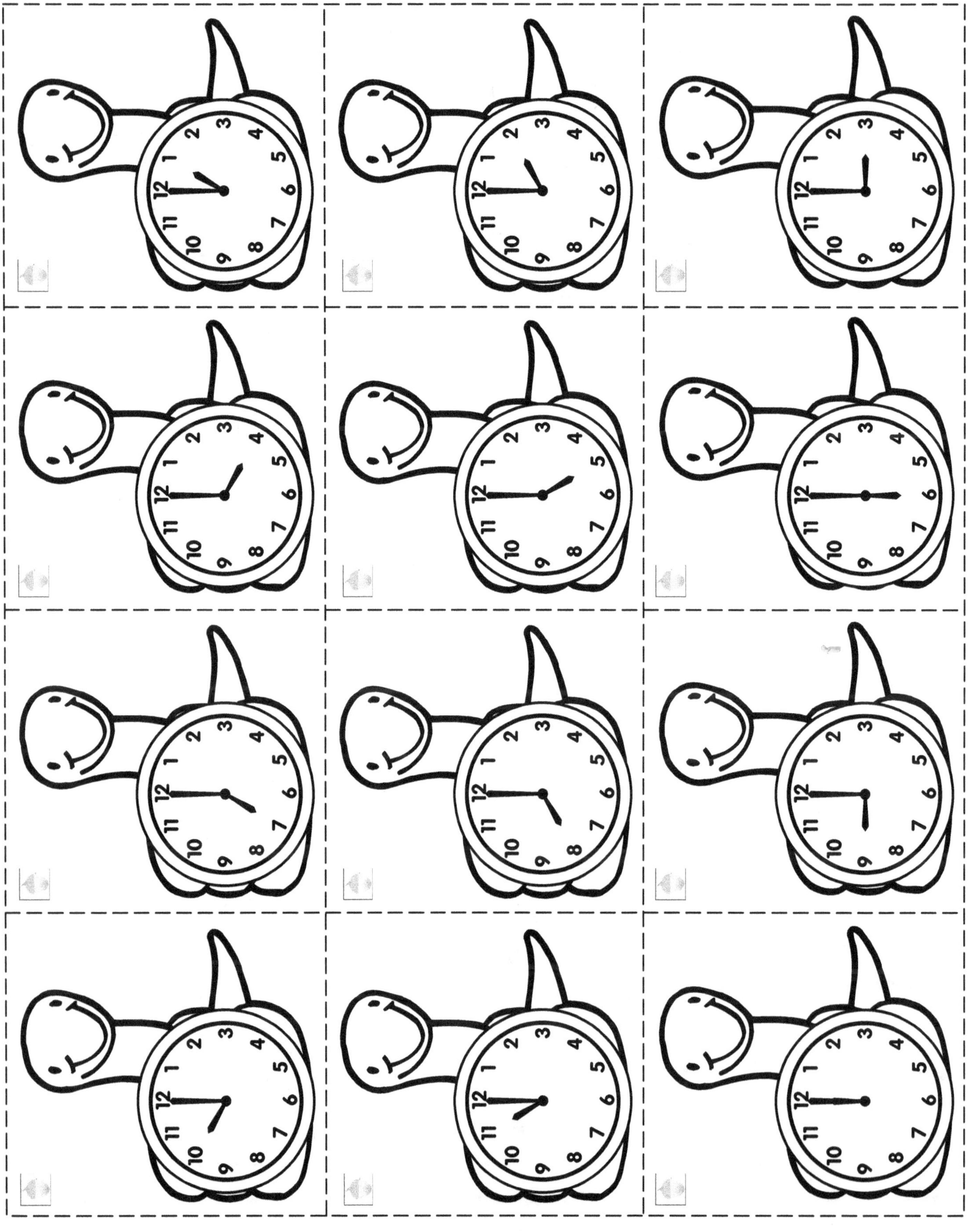

Trail Game

Sneaky Snakes Cover
Telling Time

Sneaky Snakes

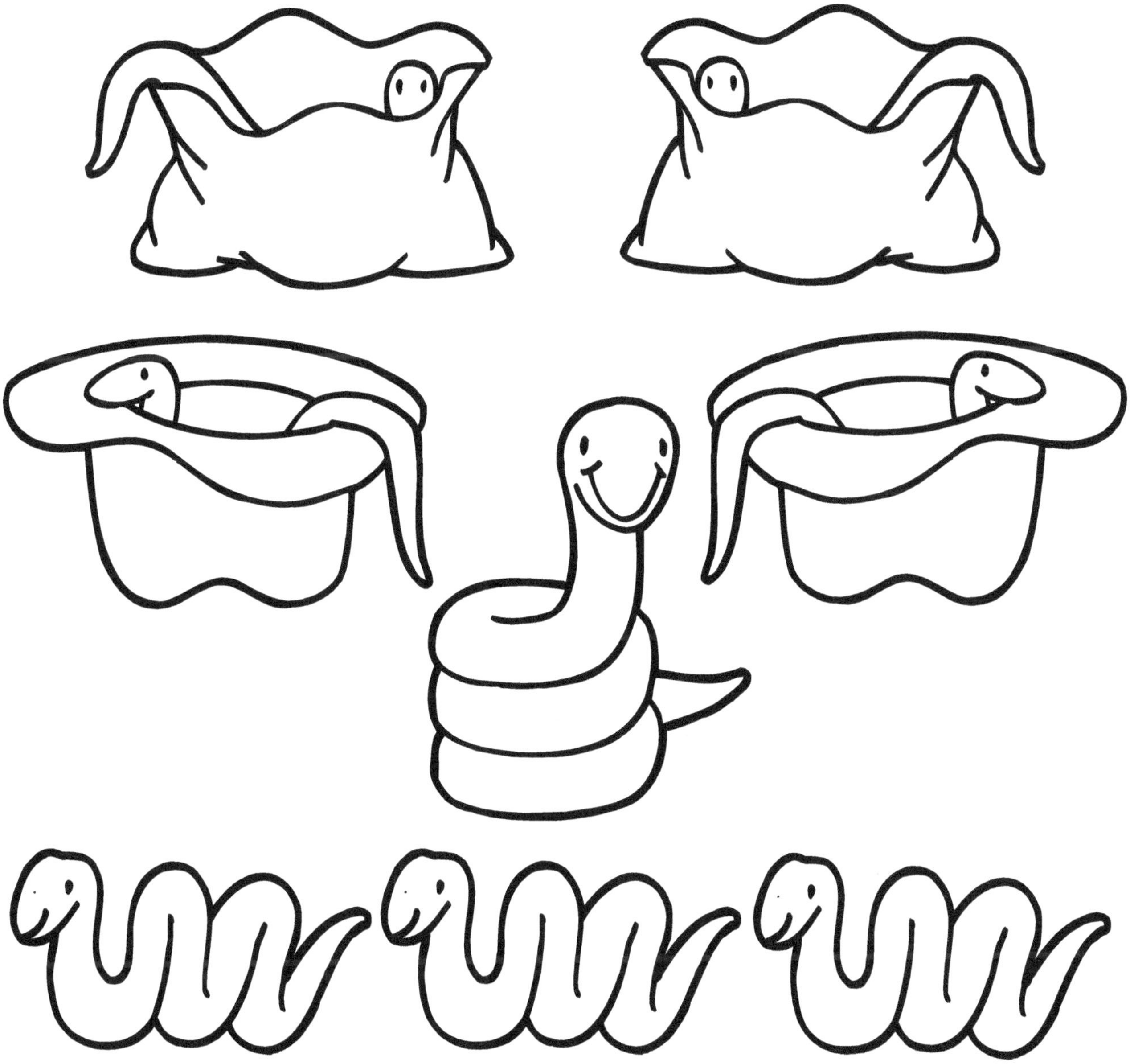

Frogs on Logs

Fill-ins

Subtracting One, Two, and Three

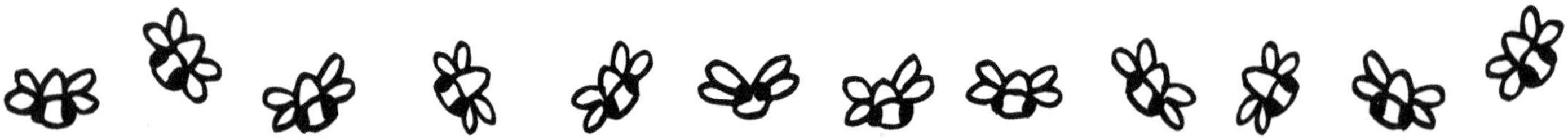

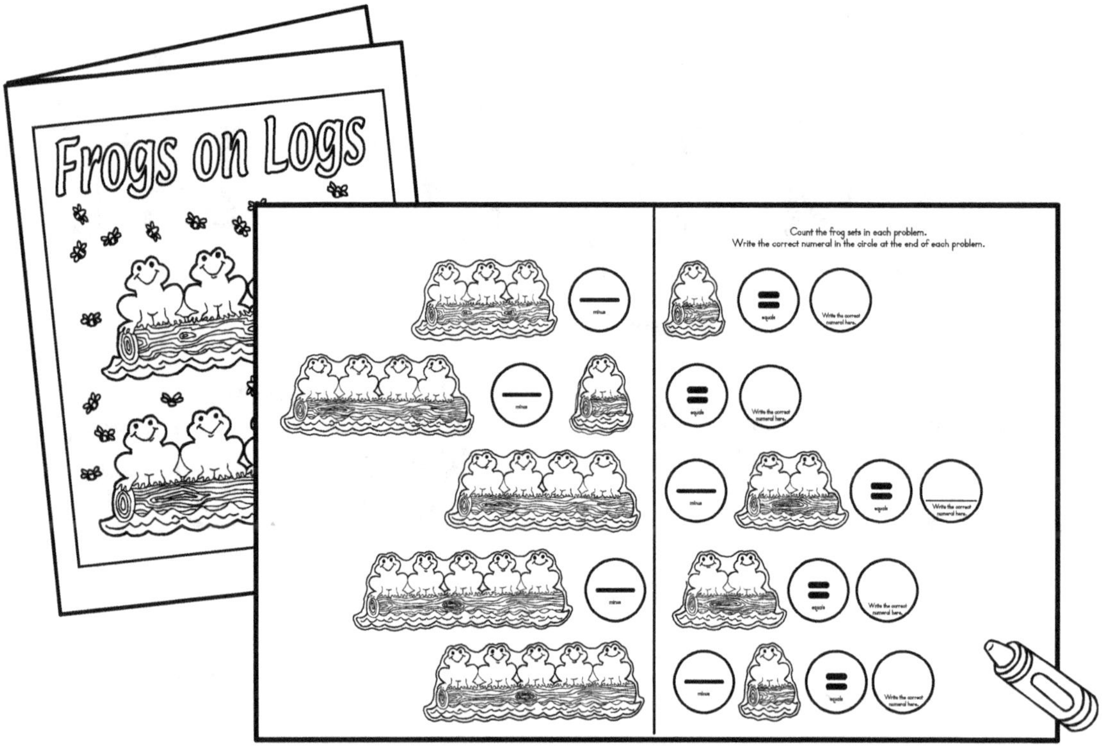

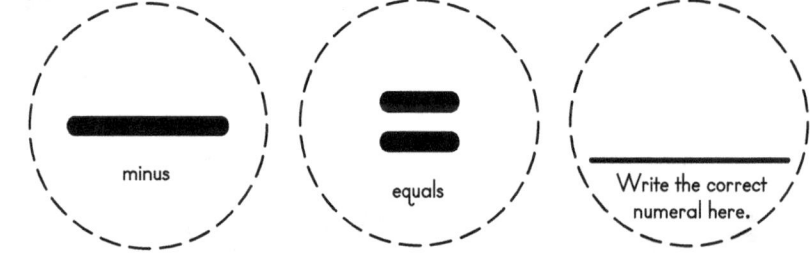

Assembly

Reproduce, cut out, and glue a set of oak tag handles (p. 192) to a standard-size file folder. Option: Decorate each handle with a copy of the flies shown above. Reproduce, color, and cut out the "Frogs on Logs" patterns. Assemble and glue two frog set outlines and symbols per problem to the inside of the folder. Glue matching frog cutouts on the frog outlines. Glue the cover art to the front of the folder. Decorate the border around the cutouts, then laminate. Glue an envelope to the back of the folder for wipe-off marker or crayon storage. Options: Reproduce, cut out, and glue the Fill-ins Playing Directions to the back of the folder.

Math Problem Symbols

Reproduce a set of symbols for each math problem.

Directions

Reproduce, cut out, and glue the direction strip to the game board.

Count the frog sets in each problem. Write the correct numeral in the circle at the end of the problem.

Frogs on Logs
Subtracting One, Two, and Three

Fill-ins

Frogs on Logs
Subtracting One, Two, and Three

Fill-ins

Reproduce, cut out, and glue two frog sets and symbols per problem to the inside of a folder.

Frogs on Logs
Subtracting One, Two, and Three

Fill-ins

Reproduce, cut out, and glue two frog sets and symbols per problem to the inside of a folder.

Shuffle Board Sharks

Adding One through Five

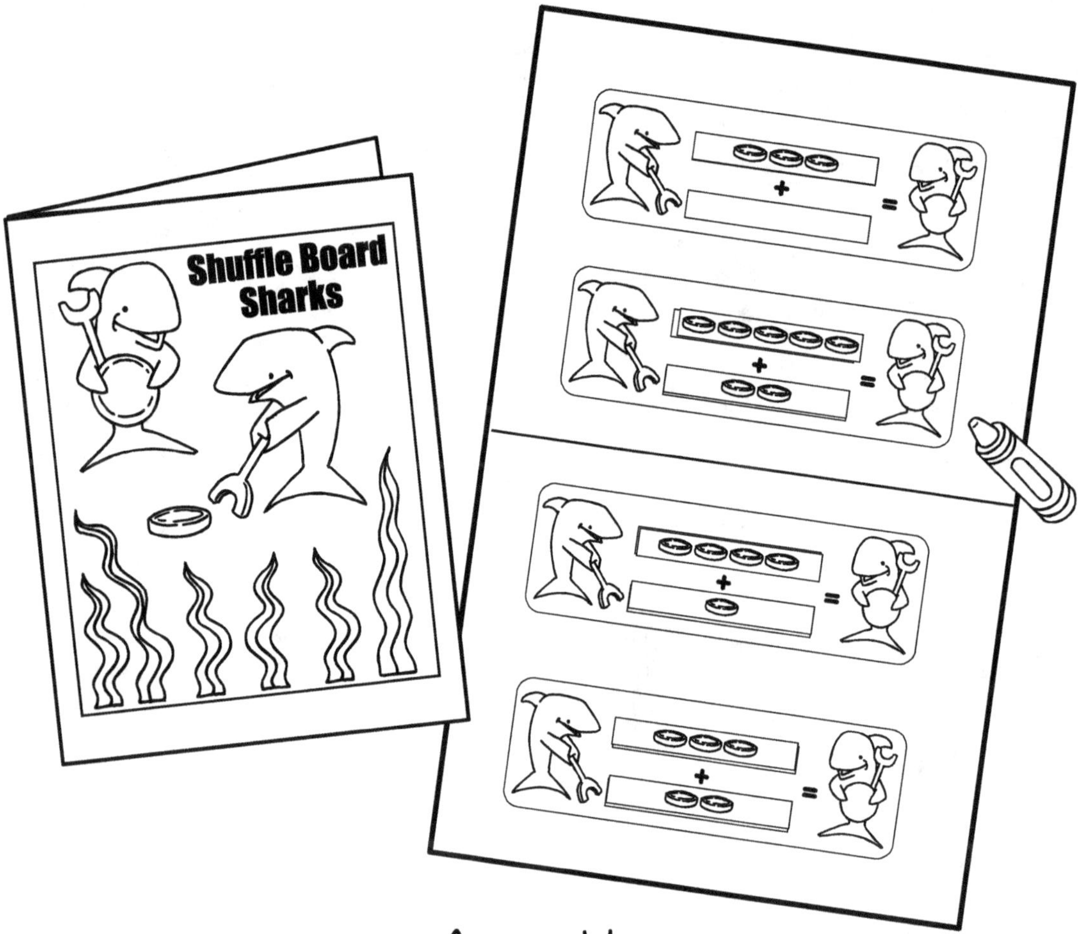

Fill-in

Assembly

Reproduce, cut out, and glue a set of oak tag handles (p. 192) to a standard-size file folder. Option: Decorate each handle with copies of the sharks shown on this page. Reproduce, color, and cut out the "Shuffle Board Sharks" patterns. Program problem sheets with puck strips, then glue programmed problem sheets to the inside of the folder. Glue the cover art to the front of the folder. Decorate the border around the fill-in boards, then laminate. Glue an envelope to the back of the folder for wipe-off marker or crayon storage. Option: Reproduce, cut out, and glue the Fill-ins Playing Directions to the back of the folder.

Shuffle Board Sharks Cover
Adding One through Five

Fill-in

Shuffle Board Sharks
Adding One through Five

Fill-in

Color, cut out, and glue four placement templates to the inside of a folder. Glue puck sets (p. 183) to form addition problems.

Count the pucks. Add the pucks. Write the correct answer on the shark's puck.

=

☐ + ☐

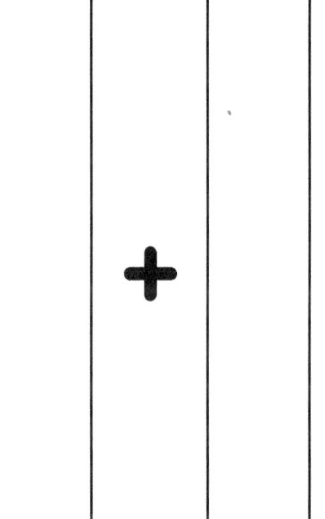

Count the pucks. Add the pucks. Write the correct answer on the shark's puck.

=

☐ + ☐

Shuffle Board Sharks
Adding One through Five

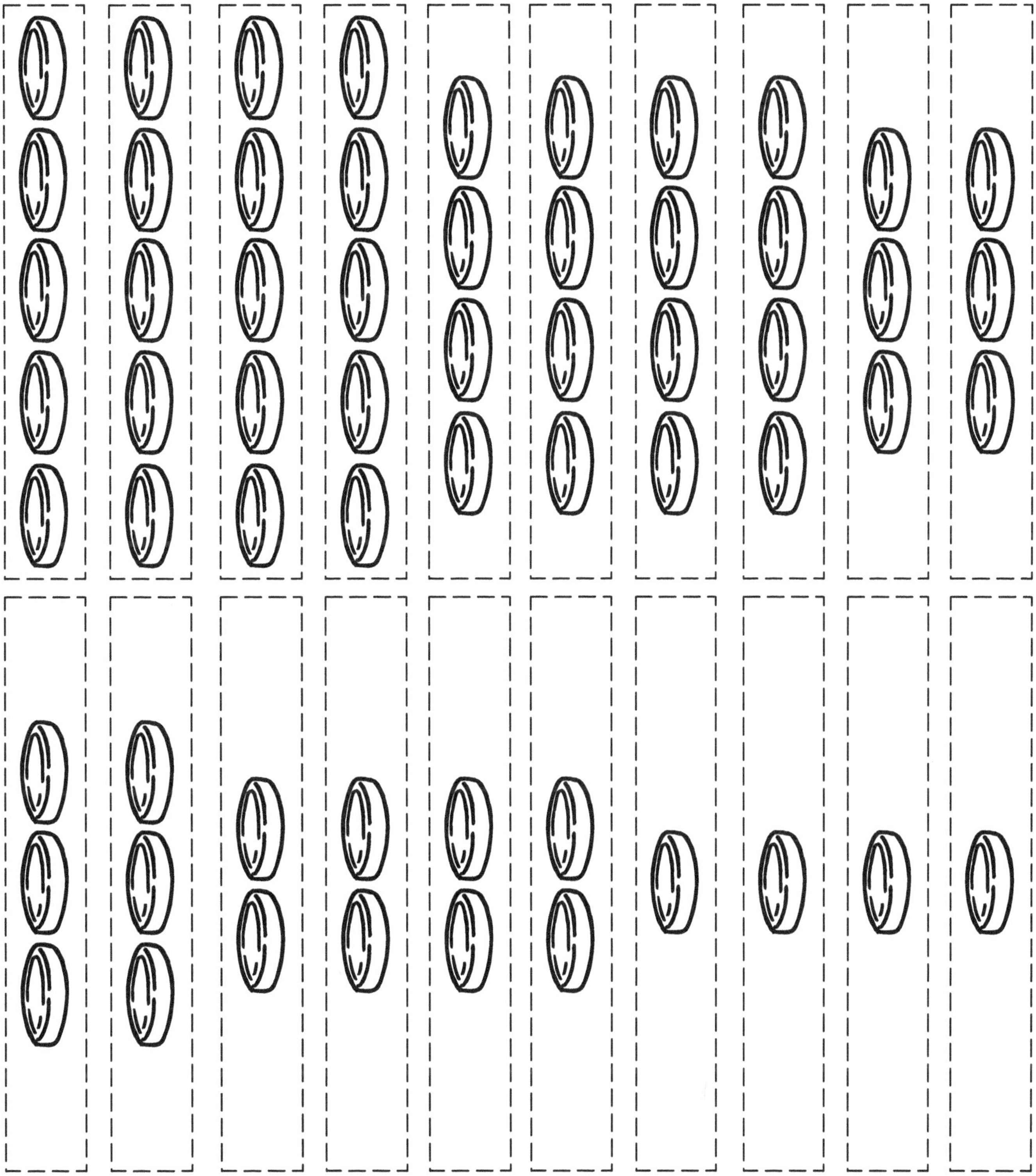

Fill-in

Whistling Whales
Counting to 100

Match Board

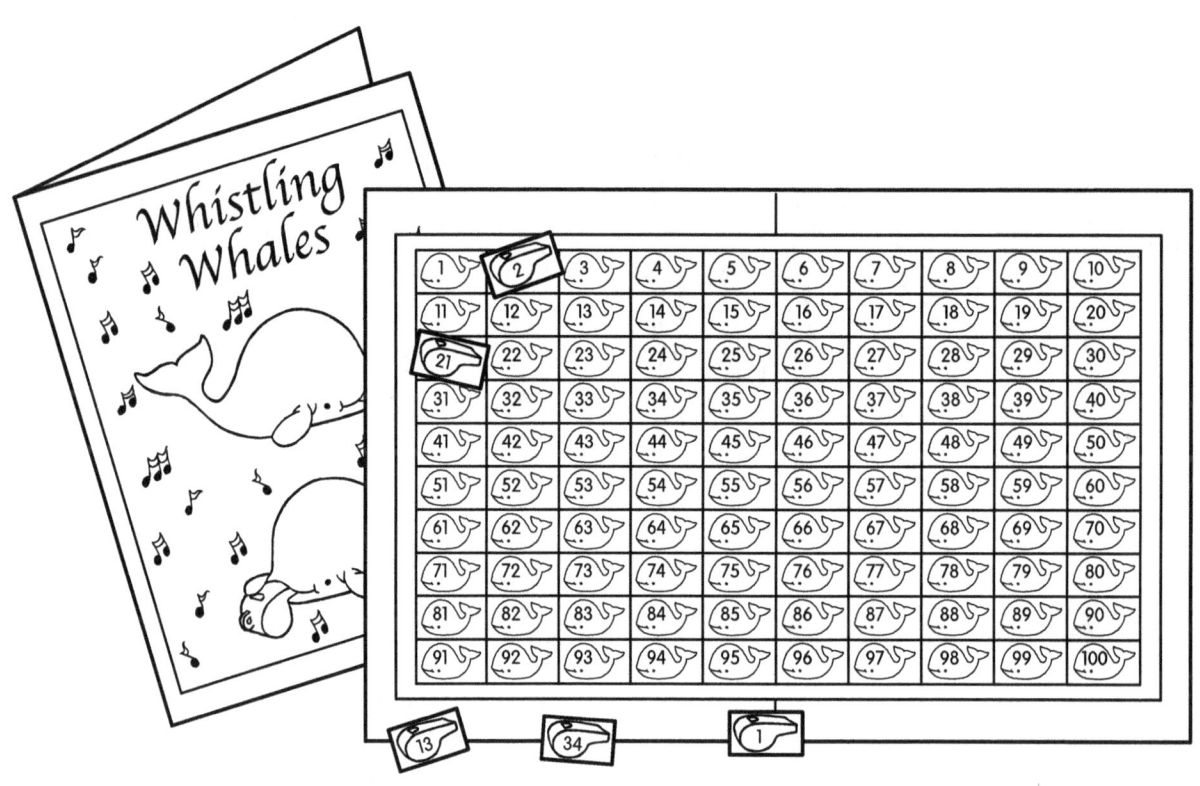

Assembly

Reproduce, cut out, and glue a set of oak tag handles (p. 192) to a standard-size file folder. Option: Decorate each handle with copies of the whales shown on this page. Reproduce, color, and cut out the "Whistling Whales" patterns. Glue the match boards to the inside of the folder. Glue the cover art to the front of the folder. Decorate the border around the match board, then laminate. Reproduce, color, laminate, then cut apart game cards. Glue an envelope to the back of the folder for game card storage. Option: Reproduce, cut out, and glue the directions (below) to the back of the folder.

Directions

Reproduce, cut out, and glue the direction strip to the game board.

Place a matching whistle on each whale.

Whistling Whales Cover
Counting to 100

Match Board

Whistling Whales

Whistling Whales
Counting to 100

Match Board

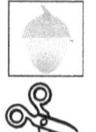

1	2	3	4	5
11	12	13	14	15
21	22	23	24	25
31	32	33	34	35
41	42	43	44	45
51	52	53	54	55
61	62	63	64	65
71	72	73	74	75
81	82	83	84	85
91	92	93	94	95

Whistling Whales

Whistling Whales
Counting to 100

Match Board

6	7	8	9	10
16	17	18	19	20
26	27	28	29	30
36	37	38	39	40
46	47	48	49	50
56	57	58	59	60
66	67	68	69	70
76	77	78	79	80
86	87	88	89	90
96	97	98	99	100

© 2014 Little Acorn Books™

Whistling Whales

Whistling Whales
Counting to 100

Match Board

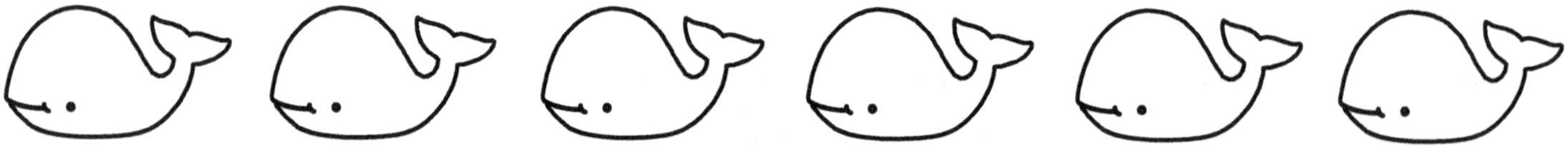

1	2	3	4	5
11	12	13	14	15
21	22	23	24	25
31	32	33	34	35
41	42	43	44	45
51	52	53	54	55
61	62	63	64	65
71	72	73	74	75
81	82	83	84	85
91	92	93	94	95

Reproduce, color, and cut apart.

Whistling Whales
Counting to 100

Match Board

6	7	8	9	10
16	17	18	19	20
26	27	28	29	30
36	37	38	39	40
46	47	48	49	50
56	57	58	59	60
66	67	68	69	70
76	77	78	79	80
86	87	88	89	90
96	97	98	99	100

Reproduce, color, and cut apart.

Whale Counters
Counting by Fives

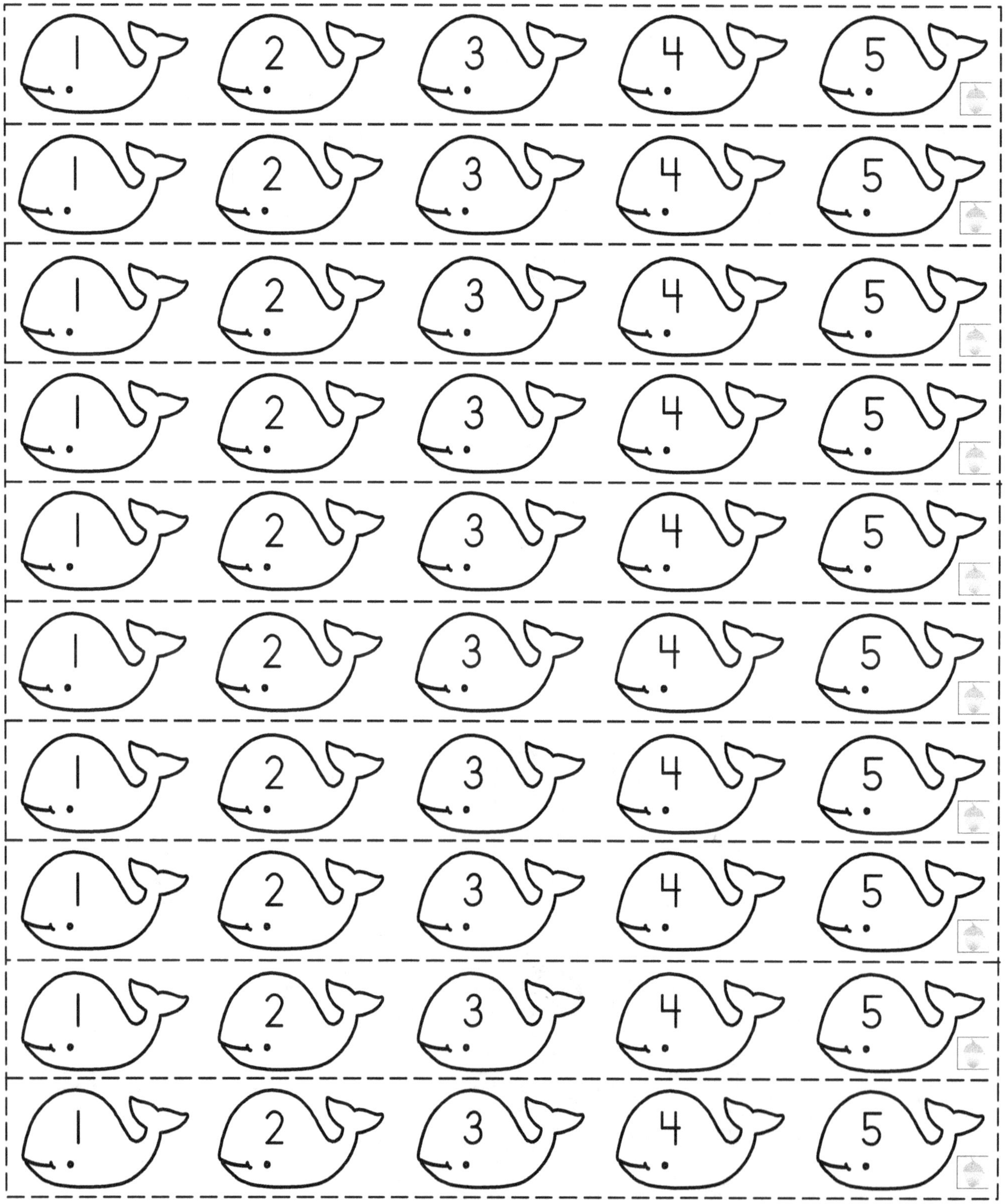

Reproduce, color, and cut apart two sets of strips for children to practice counting to 100 by fives.

Whale Counters
Counting by Tens

Reproduce, color, and cut apart two sets of strips for children to practice counting to 100 by tens.

Folder Handles

Reproduce, color, and cut out a set of handles for each folder game.

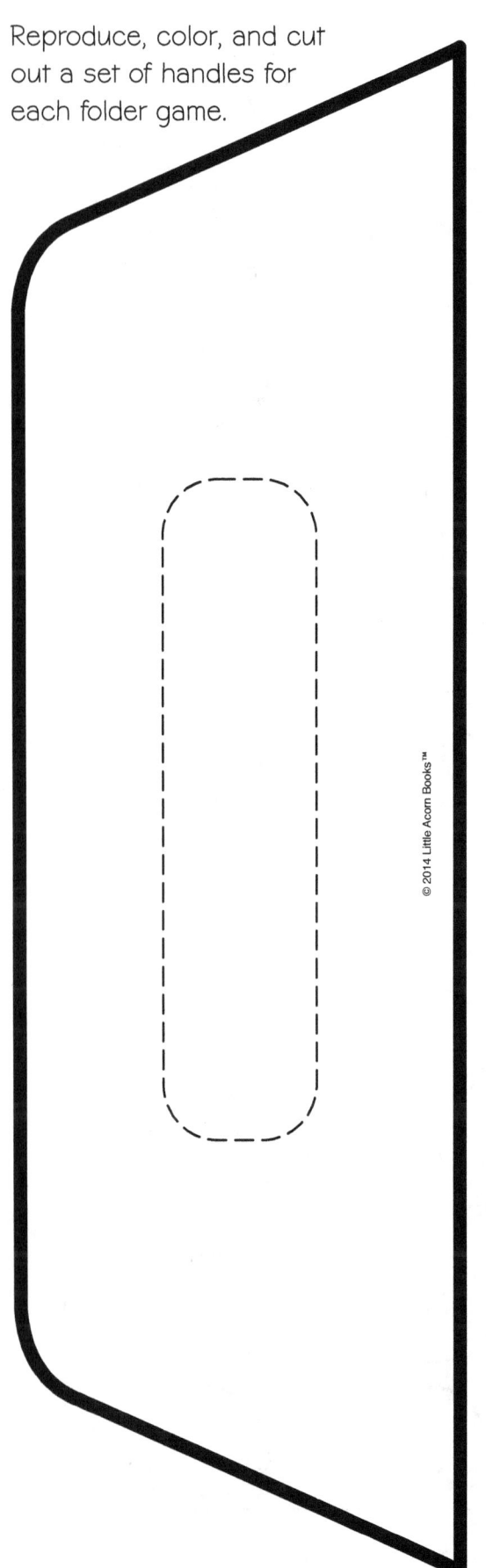
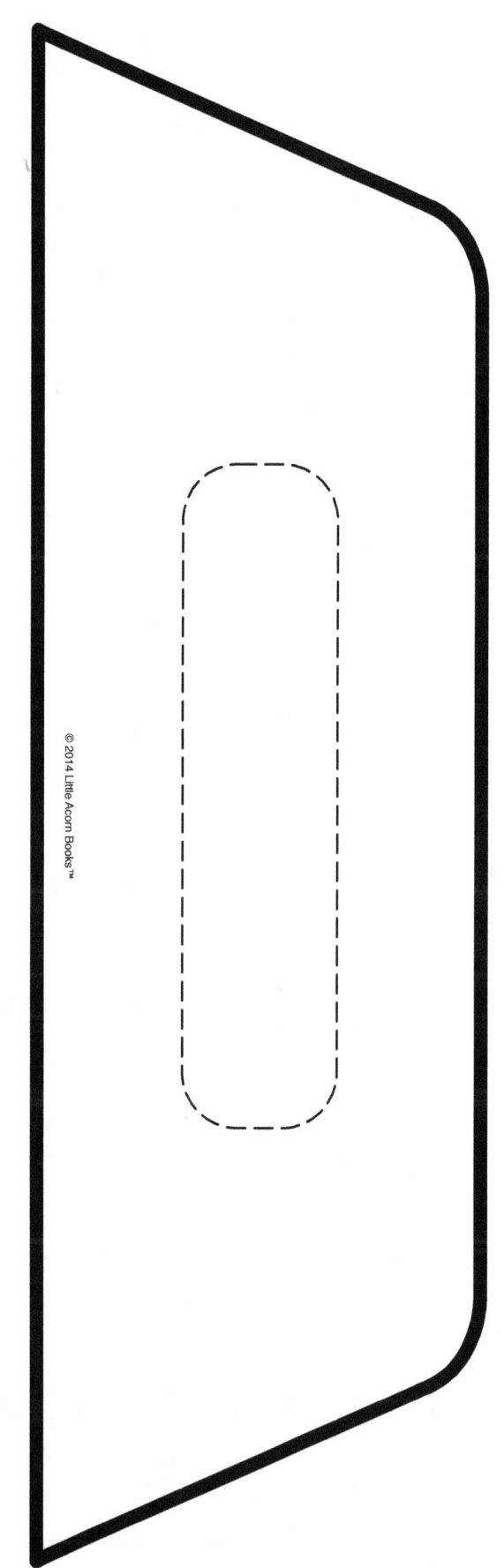

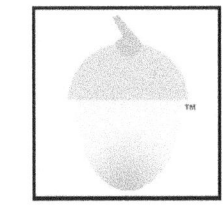

Little Acorn Books™

Promoting Early Skills for a Lifetime™

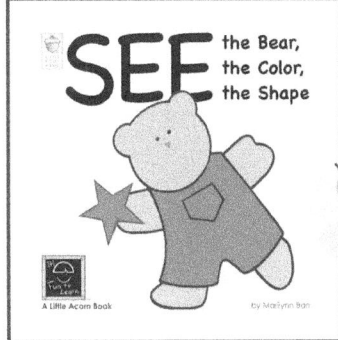   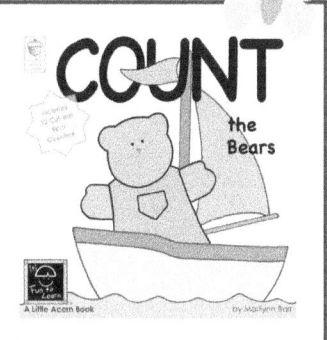

A Hands-on Picture Book Series • Infancy–Age 4

Miss Pitty Pat & Friends
Preschool–Grade 1

Using Crayons, Scissors, & Glue for Crafts
Preschool–Grade 1

Mookie's Christmas Tree
For All Ages and Not Just for Christmas

Little Acorn Books™
Visit our web site:
www.littleacornbooks.com

www.ingramcontent.com/pod-product-compliance
Lightning Source LLC
Chambersburg PA
CBHW081456040426
42446CB00016B/3264